CMP

Advances in Boundary Elements

Vol.1: Computations and Fundamentals

Proceedings of the eleventh International Conference on Boundary Element Methods, Cambridge, USA, August 1989

Editors:
C.A. Brebbia
J.J. Connor

Computational Mechanics Publications
Southampton Boston

Springer-Verlag Berlin Heidelberg
New York London Paris Tokyo

Carlos A. Brebbia
Computational Mechanics Institute
Ashurst Lodge
Ashurst
Southampton
SO4 2AA
UK

Jerome J. Connor
Dept. of Civil Engineering
Massachusetts Institute of Technology
Cambridge
MA 02139
USA

British Library Cataloguing in Publication Data
International Conference on Boundary Element Methods
(11th : 1989 : Cambridge, Mass.)
Advances in boundary elements.
1. Engineering, Mathematics. Boundary element methods
I. Title II. Brebbia, C. A. (Carlos Alberto), *1938-*
III. Connor, J. J. (Jerome Joseph), *1932-*
620'.001'515353
ISBN 1-85312-024-3
ISBN 1-85312-025-1
ISBN 1-85312-026-X

Library of Congress Catalog Card Number 89-61999

ISBN 1-85312-024-3	Computational Mechanics Publications Southampton
ISBN 0-945824-10-6	Computational Mechanics Publications Boston
ISBN 3-540-51448-1	Springer-Verlag Berlin Heidelberg New York London Paris Tokyo
ISBN 0-387-51448-1	Springer-Verlag New York Heidelberg Berlin London Paris Tokyo
ISBN 1-85312-027-8	
ISBN 3-540-51452-X	3 volume set
ISBN 0-387-51452-X	

Printed and bound by Bookcraft (Bath) Ltd., Avon.

CONTENTS

SECTION 4 - COMPUTATIONAL ASPECTS

PREFACE

Since its humble beginnings in 1978, when the name 'Boundary Elements' was used for the first time, the International Conference on Boundary Element Methods in Engineering has grown into a major annual event, during which the more recent developments of the method are presented, long before they are published in any scientific journal. The meeting has become the focus for BEM research and provides a unique forum for the international scientific community to gather and discuss problems of current interest.

It is to us, the editors of these volumes, a source of particular pride that the Conference appeals to comparatively young researchers as well as those already well established. It has also become a remarkable meeting for its friendly atmosphere and has contributed in no small part to creating permanent links between researchers from many different countries.

This book contains some of the edited papers presented at the 11th Boundary Element Conference, held in Cambridge, Massachusetts, during August 1989. The papers are arranged in three different books comprising the following topics:

Vol. 1 **Computations and Fundamentals** - comprises sections on fundamentals, adaptive techniques, error and convergence, numerical methods and computational aspects.

Vol. 2 **Field and Fluid Flow Solutions** - includes the following topics: potential problems, thermal studies, electrical and electromagnetic problems, wave propagation, acoustics and fluid flow.

Vol. 3 **Stress Analysis** - deals with advances in linear problems, nonlinear problems, fracture mechanics, contact mechanics, optimization, geomechanics, plates and shells, vibrations and industrial applications.

The next Conferences will be held in the following locations:

BEM 12 in Hokkaido, Japan in 1990
BEM 13 in Tulsa, USA, in 1991
BEM 14 in Seville, Spain in 1992

Further information about this series can be obtained from the sponsors, i.e.

International Society for Boundary Elements (ISBE)
Ashurst Lodge
Ashurst, Southampton SO4 2AA
Tel 44-(0)703-293223
Fax 44-(0)703 292853

The Editors are indebted to all contributors who with the quality of their papers have maintained the high standards which are the trademark of this series of Conferences.

Carlos A. Brebbia
Jerome J. Connor

SECTION 1 - FUNDAMENTALS

Application of the Generalized Boundary Element Method to Nonlinear Problems

Nobuyoshi Tosaka
Dept. of Mathematical Engineering, College of Industrial Technology, Nihon University, Narashino, Chiba, 275, Japan

INTRODUCTION

According to recent notable development of the boundary element method, we now encounter the question of applicability and adaptability of the method to nonlinear problems in various fields. In order to answer this question which has been mainly posed by the finite element researchers we must study on some expansion of the boundary element method and propose several effective procedures.

We have been challenging to give the effectiveness and adaptability of the integral equation method through applications to many kinds of nonlinear problems. In the early years of our study, we proposed the boundary-domain element approach and applied this approach to geometrically nonlinear problems of thin elastic body, e.g. Tosaka and Miyake [1] and viscous fluid flow problems, e.g. Tosaka [2]. Recently, we proposed a new approach which we called the generalized boundary element approach, e.g. Tosaka and Kakuda [3], Tosaka [4]. In References [3,4], we show mainly the effectiveness and applicability of the new approach in one-dimensional nonlinear problems.

In this paper, we would like to discuss an attempt to extend applicability of the generalized boundary element method. After a short explanation of the method, numerical examples of the two-dimensional nonlinear problem are presented in order to illustrate the adaptability and effectiveness of the approach. They are the nonlinear chemical reaction problem and the Plauteu's problem known as the minimum surface problem in differential geometry.

GENERALIZED BOUNDARY ELEMENT METHOD

Let Ω be a domain in $\mathbf{R}^2$ and Γ be a boundary curve with the unit normal vector n_i . In this geometrical region we will be concerned with integral equation formulations of the following boundary-value problem of a nonlinear differential equation:

$$A(u) = Lu + N(u) = f \qquad \text{in } \Omega \tag{1}$$

$$u = u \qquad \text{on } \Gamma_u \tag{2}$$

$$Bu = q = q \qquad \text{on } \Gamma_q \tag{3}$$

where A is a nonlinear differential operator defined on the functional space on Ω and is given by the sum of a linear part L and a nonlinear part N, B denotes a boundary operator related to A, and $\hat{u}$ and $\hat{q}$ are the Dirichlet and Neumann data, respectively. Equation (1) is an abstract form of the fundamental equations of science and engineering.

There exist several approaches to determine the solution of the problem (1)-(3) by using the integral equation method. In Reference [3], I proposed typical three approaches which are the boundary element approach, the boundary-domain element approach and the generalized boundary element approach. These approaches have mutual distinction according to the fundamental solution adopted in derivation of each integral equation formulation.

The essential idea and numerical solution procedure in the generalized approach are as follows:

1. The linearization of nonlinear equation

The linearization of nonlinear equation (1) is performed in subdomain ${}_i\Omega$ of the given domain Ω . The linearized expression of (1) can be given under the linearization of nonlinear operator, N(u)=vMu , as

$$Lu + vMu = f \qquad \text{in } {}_i\Omega \tag{4}$$

where M denotes the linear operator.

2. Derivation of boundary integral equation

We can easily derive the boundary integral equation of the differential equation (4) as

$$
\begin{aligned}
cu(\mathbf{y}) = & \int_{{}_i\Gamma} (B^* w^*(\mathbf{x},\mathbf{y})) u(\mathbf{x}) d\Gamma(\mathbf{x}) - \int_{{}_i\Gamma} (Bu(\mathbf{x})) w^*(\mathbf{x},\mathbf{y}) d\Gamma(\mathbf{x}) \\
& + \int_{{}_i\Gamma} v\{(B^* w^*(\mathbf{x},\mathbf{y})) u(\mathbf{x}) - (Bu(\mathbf{x})) w^*(\mathbf{x},\mathbf{y})\} d\Gamma(\mathbf{x}) \\
& + \int_{{}_i\Omega} f(\mathbf{x}) w^*(\mathbf{x},\mathbf{y}) d\Omega(\mathbf{x}) \qquad (5)
\end{aligned}
$$

where B is the boundary operator associated with M and $w^*(\mathbf{x},\mathbf{y})$ is the fundamental solution for the adjoint of operator (L+vM) which satisfies

$$(L+vM)^* w^*(\mathbf{x},\mathbf{y}) = \delta(\mathbf{x}-\mathbf{y}) \qquad (6)$$

3. Discretization with boundary elements.

Using the standard boundary element technique, the discretized form of (5) is given as follows:

$${}_i\mathbf{H}\,{}_i\mathbf{u} = {}_i\mathbf{G}\,{}_i q + {}_i\mathbf{f} \qquad (7)$$

where ${}_i\mathbf{H}$ and ${}_i\mathbf{G}$ are the influence matrices, ${}_i\mathbf{u}$ and ${}_i\mathbf{q}$, are the unknown nodal vectors and ${}_i\mathbf{f}$ is the given vector.

4. Combination of discretized forms

Taking account of connectivity conditions on the interface between subdomains, we can connect the discretized forms with each other and set the whole system as

$$\mathbf{HU} = \mathbf{GQ} + \mathbf{B} \qquad (8)$$

5. Setting up the final system

Putting the boundary conditions into (8) and reordering the unknowns of the problem, we can set up the final quasi-linear system as follows:

$$\mathbf{A}(\mathbf{v})\mathbf{X} = \mathbf{F} \qquad (9)$$

where $\mathbf{A}(\mathbf{v})$ denotes the coefficient matrix which depends on the assumed value $\mathbf{v}$, $\mathbf{X}$ is the unknown vector and $\mathbf{F}$ is the known vector.

6. Solution procedure

We can solve the quasi-linear system (9) by using the simple iterative solution procedure.

NUMERICAL EXAMPLES

In the following, two examples on the two-dimensional problem are presented in order to illustrate the proposed method.

1. Chemical Reaction Problem

Let us consider the following boundary-value problem in chemical engineering:

$$\Delta u = \alpha u^2 \qquad \text{in } \Omega \tag{10}$$

$$u(0,y) = 1.0, \; u(1,y) = 0.25 \tag{11}$$

$$\frac{\partial u}{\partial n}(x,0) = \frac{\partial u}{\partial n}(x,1) = 0 \tag{12}$$

where α is the positive constant. The Dirichlet problem of Eq.(10) arises in obtaining the concentration u of the species in diffusion supported chemical reaction proceeding at constant temperature. Three iterative processes yielding the solution were presented in Reference [6].

In this paper, we would like to solve approximately the above problem by means of the generalized boundary element approach. First of all, we divide the whole domain Ω into N subdomains such that $\Omega = \bigcup_{i=1}^{N} {}_i\Omega$ and linearize equation (10) on the ith subdomain, ${}_i\Omega$, as follows:

$$\Delta u = \alpha v u \qquad \text{in } {}_i\Omega \tag{13}$$

in which v is assumed to be the known constant in ${}_i\Omega$. Because this equation is linear, we can easily derive the following boundary integral equation of (13):

$$c(\mathbf{y})u(\mathbf{y}) = \int_{{}_i\Gamma} u(\mathbf{x}) \frac{\partial w^*(\mathbf{x},\mathbf{y})}{\partial n} d\Gamma(\mathbf{x}) - \int_{{}_i\Gamma} \frac{\partial u(\mathbf{x})}{\partial n} w^*(\mathbf{x},\mathbf{y}) d\Gamma(\mathbf{x}) \tag{14}$$

where $c(\mathbf{y})$ denotes the so-called shape coefficient and w^* is the fundamental solution for the modified Helmholtz differential operator which satisfies

$$(\Delta - \alpha v) w^* = \delta(\mathbf{x}-\mathbf{y}) \tag{15}$$

We can determine the fundamental solution and its derivative as follows:

$$w^*(\mathbf{x},\mathbf{y}) = -\frac{1}{2\pi} K_0(\sqrt{\alpha v}\, r) \tag{16}$$

$$\frac{\partial w^*}{\partial n}(\mathbf{x},\mathbf{y}) = \frac{\sqrt{\alpha v}}{2\pi} K_1(\sqrt{\alpha v}\, r) \frac{(\mathbf{x}-\mathbf{y})\cdot \mathbf{n}}{r} \tag{17}$$

where K_ν (ν =0,1) denotes the modified Bessel function and r= $\| \mathbf{x}-\mathbf{y} \|$.

Following the standard manner of discretization in the boundary element method, we obtain the discretized expression of (14) for any subdomain $_i\Omega$ with the boundary $_i\Gamma$ in matrix form as

$$ {}_i\mathbf{H}\,{}_i\mathbf{u} = {}_i\mathbf{G}\,{}_i\mathbf{q} \tag{18} $$

where the matrices $_i\mathbf{H}$ and $_i\mathbf{G}$ are influence matrices, and $_i\mathbf{u}$ and $_i\mathbf{q}$ are the nodal vectors. Taking into consideration the continuity conditions on the interface boundary in setting up (8) for all subdomains, we can obtain the final system in terms of the unknown vector $\mathbf{X}$ as follows:

$$ \mathbf{A}(\mathbf{v})\mathbf{X} = \mathbf{F} \tag{19} $$

where $\mathbf{v}$ denotes the vector of values assumed in each subdomain in advance and $\mathbf{A}(\mathbf{v})$ is the coefficient matrix dependent on $\mathbf{v}$.

Let us show the numerical results obtained for cases of several values of the constant α. The element meshes used in this paper are shown in Fig.1.

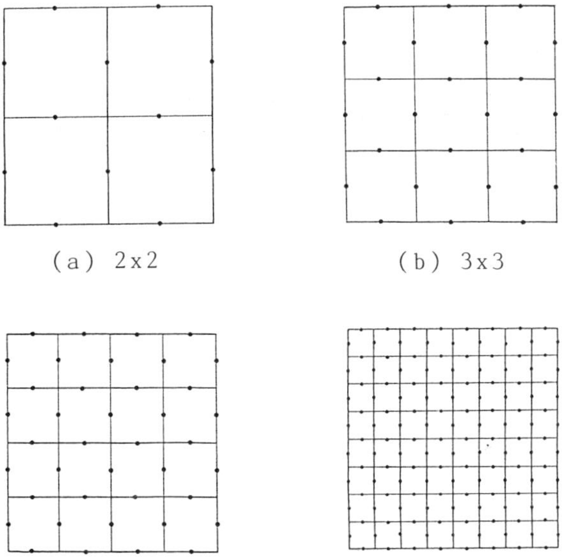

(a) 2x2 (b) 3x3 (c) 4x4 (d) 8x8

Fig.1 Element meshes

Fig.2 shows the numerical results in the case of $\alpha=6$ for each number of subdomains and comparison of these with the exact solution given by

$$u(x,y) = \frac{1}{(1+x)^2}, \quad u_{,x}(x,y) = \frac{-2}{(1+x)^3} \tag{20}$$

In this case, our results indicate a good agreement with the above analytical solution.

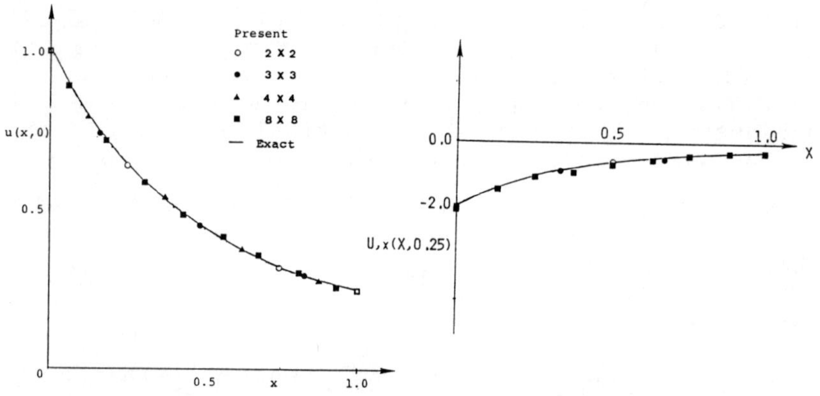

(a) u(x,0) (b) $u_{,x}(x,0.25)$

Fig.2 Numerical results ($\alpha=6$)

Figs.3 and 4 show the numerical results for larger values of α. It is obvious that the method leads to reasonable numerical results for the problems with large constant α.

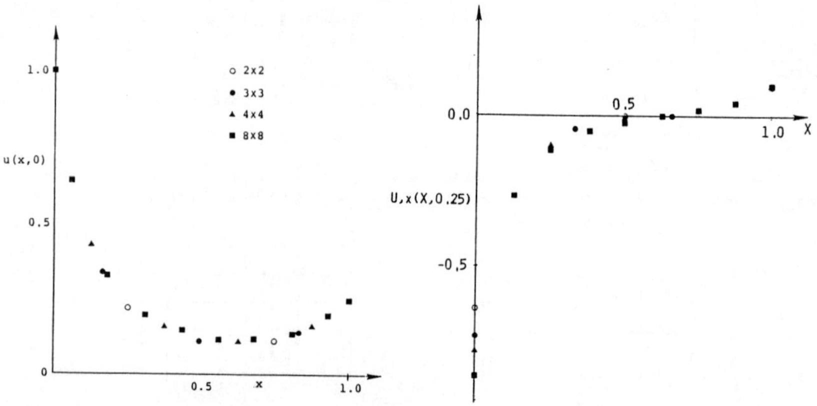

(a) u(x,0) (b) $u_{,x}(x,0.25)$

Fig.3 Numerical results ($\alpha=10^2$)

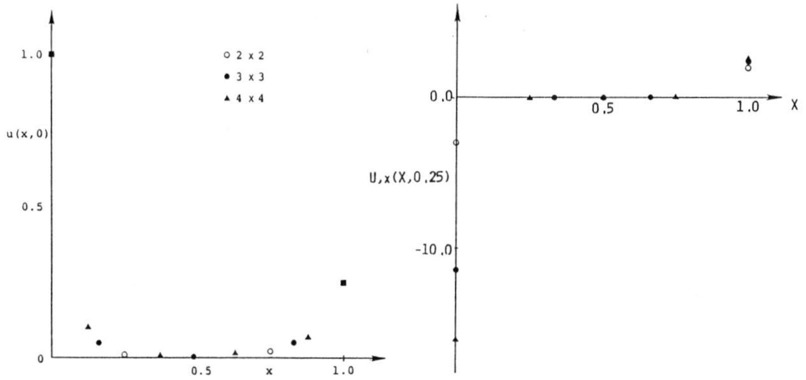

(a) $u(x,0)$ (b) $u_{,x}(x,0.25)$

Fig.4 Numerical results ($\alpha=10^3$)

2. Plateau's Problem

Next, let us consider application to Plateau's problem which is one of the well-known problems in differential geometry. The problem is to determine the surface of minimum area in three dimensional Euclidean space for the given boundary spacial curve of it. Usually, this problem is formulated with the variational form. The well-known Euler equation of this variational formulation is given by the following higher (i.e. the fourth order) nonlinear equation in terms of the surface $z=u(x,y)$:

$$\frac{\partial^2 u}{\partial x^2}\{1+(\frac{\partial u}{\partial y})^2\}-2\frac{\partial^2 u}{\partial x\partial y}\frac{\partial u}{\partial x}\frac{\partial u}{\partial y}+\frac{\partial^2 u}{\partial y^2}\{1+(\frac{\partial u}{\partial x})^2\}=0 \quad \text{in } \Omega \qquad (21)$$

The boundary condition is also given by

$$u(x,y) = g(x,y) \qquad \text{on } (x,y)\in\Gamma\equiv\partial\Omega \qquad (22)$$

in which g is assumed to be a given single-valued function.

Following the solution procedure in Reference [7], we start our analysis with the following linearized equation in any subdomain ${}_i\Omega$:

$$(1+\beta^2)\frac{\partial^2 u}{\partial x^2}-2\alpha\beta\frac{\partial^2 u}{\partial x\partial y}+(1+\alpha^2)\frac{\partial^2 u}{\partial y^2}=0 \qquad \text{in } {}_i\Omega \qquad (23)$$

where $\alpha(=\frac{\partial u}{\partial x})$ and $\beta(=\frac{\partial u}{\partial y})$ are assumed to be constant in ${}_i\Omega$. The boundary integral equation of (23) can be given as

$$c(\mathbf{y})u(\mathbf{y})=\int_{{}_i\Gamma}\{u(\mathbf{x})q^*(\mathbf{x},\mathbf{y})-q(\mathbf{x})w^*(\mathbf{x},\mathbf{y})\}d\Gamma(\mathbf{x})$$

$$+\int_{{}_i\Gamma}\{u(\mathbf{x})Q^*(\mathbf{x},\mathbf{y})-Q(\mathbf{x})w^*(\mathbf{x},\mathbf{y})\}d\Gamma(\mathbf{x}) \quad (24)$$

in which we introduce the following quantities,

$$q(\mathbf{x}) = \frac{\partial u}{\partial n}, \quad q^*(\mathbf{x},\mathbf{y}) = \frac{\partial w^*}{\partial n} \quad (25)$$

$$\left.\begin{aligned} Q(\mathbf{x})&=\beta^2\frac{\partial u}{\partial x}n_x-\alpha\beta(\frac{\partial u}{\partial x}n_y+\frac{\partial u}{\partial y}n_x)+\alpha^2\frac{\partial u}{\partial y}n_y \\ Q^*(\mathbf{x},\mathbf{y})&=\beta^2\frac{\partial w^*}{\partial x}n_x-\alpha\beta(\frac{\partial w^*}{\partial x}n_y+\frac{\partial w^*}{\partial y}n_y)+\alpha^2\frac{\partial w^*}{\partial y}n_y \end{aligned}\right\} \quad (26)$$

The fundamental solution w* which should satisfy

$$(1+\beta^2)\frac{\partial^2 w^*}{\partial x^2}-2\alpha\beta\frac{\partial^2 w^*}{\partial x\partial y}+(1+\alpha^2)\frac{\partial^2 w^*}{\partial y^2}=\delta(\mathbf{x}-\mathbf{y}) \quad (27)$$

is now given by the following logarithmic expression:

$$w^*(\mathbf{x},\mathbf{y}) = \frac{1}{2\pi\hat{\alpha}\hat{\beta}\sqrt{1-\gamma 2}}\ln R \quad (28)$$

where

$$\hat{\alpha}= 1+\alpha^2, \quad \hat{\beta}= 1+\beta^2, \quad \gamma=\frac{\alpha\beta}{\hat{\alpha}\hat{\beta}} \quad (29)$$

$$R^2=\frac{1}{2(1-\gamma)}(\frac{x_1-y_1}{\hat{\beta}}+\frac{x_2-y_2}{\hat{\alpha}})^2+\frac{1}{2(1+\gamma)}(\frac{x_1-y_1}{\hat{\beta}}-\frac{x_2-y_2}{\hat{\alpha}})^2 \quad (30)$$

The above boundary integral equation (24) can be discretized and then solved approximately in a similar way to the previous problem.

Let us show the minimum surfaces with a plane form calculated by means of our approach. The number of elements used in this performance is 20x20.

Fig.5 shows the obtained minimum surface with a boundary curve of four linear edges. It is obvious that the surface looks like the hyperbolic-paraboloid. The obtained numerical results for the surface with a boundary curve of two linear edges and two circular edges are also shown from two viewpoints in Fig.6.

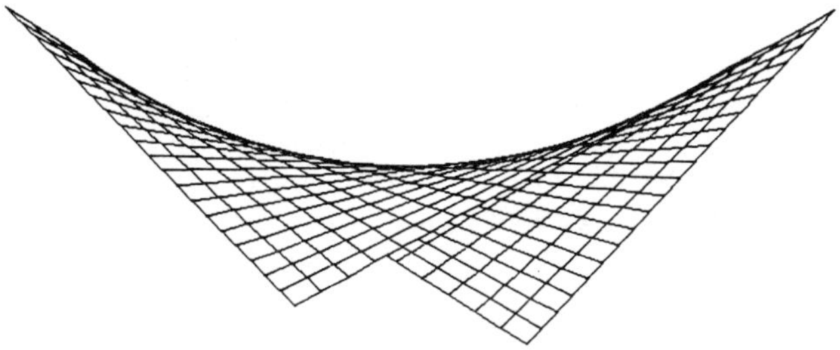

Fig.5 Minimum surface with four linear edges

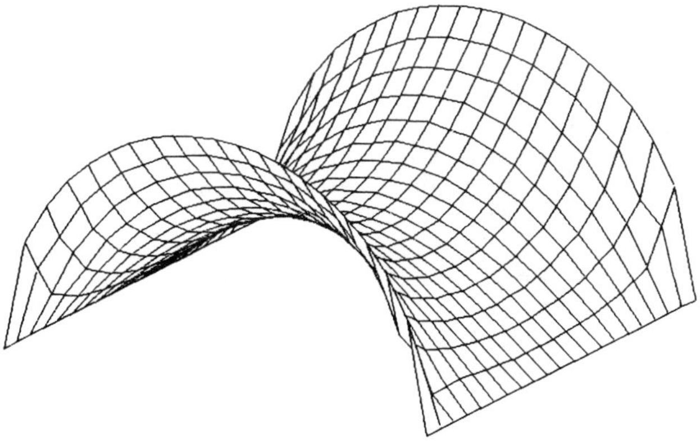

(a)

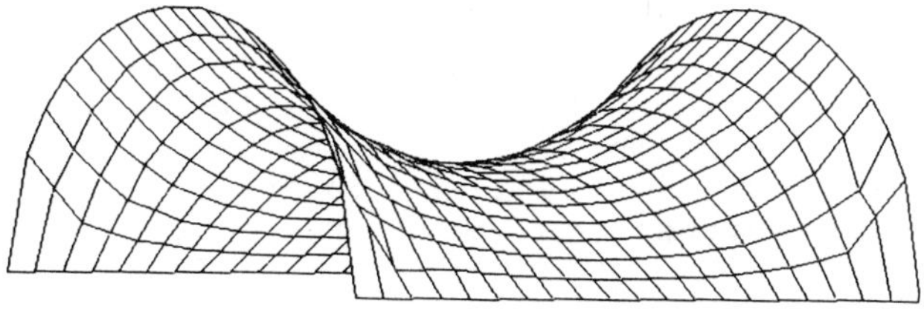

(b)

Fig.6 Minimum surface with two linear and two circular edges

CONCLUSIONS

In this paper, applicability of the generalized boundary element approach in the integral equation method to two-dimensional nonlinear problems is mainly discussed. The accuracy of this approach is demonstrated in analyzing two nonlinear problems which are the chemical reaction problem and the Plateau's problem. The procedure leads to reasonable numerical results for cases with large constant in the chemical reaction problem. Moreover, the method is also effective for the highly nonlinear problem in which fourth order nonlinear term is included. It is evident that our approach is very effective for not only the one-dimensional problem but also the two-dimensional one.

REFERENCES

1. Tosaka, N. and Miyake, S. Integral Equation Analysis for Geometrically Nonlinear Problems of Elastic Bodies, Theory and Applications of Boundary Element Methods (Eds. Tanaka M. and Du Qinghua), Pergamon Press, pp.251-260, 1987.
2. Tosaka, N. Integral Equation Formulations with the Primitive Variables for Incompressible Viscous Fluid Flow Problems, Computational Mechanics, Vol.4, pp.89-103, 1989.
3. Tosaka, N. and Kakuda, K. The Generalized Boundary Element Method for Nonlinear Problems, in Boundary Elements X (Ed. Brebbia, C.A.), Vol.1, pp.3-17, Springer-Verlag, 1988.
4. Tosaka, N. Boundary Element Method and Nonlinear Problems, in Boundary Element Method in Applied Mechanics (Eds. Tanaka, M. and Cruse, T.A.), Pergamon Press, pp.491-500, 1988.
5. Tosaka, N. Nonlinear Analysis, Topics in Boundary Element Research (Ed. Brebbia, C. A.) Vol.8, (to appear).
6. Ablow, C. M. and Perry, C. L. Iterative Solutions of the Dirichlet Problem for $\Delta u=u^2$, J. Soc. Indust. Appl. Math., Vol.7, pp.459-467, 1959.
7. Tosaka, N. and Fukushima, N. Generalized Boundary Element Approach to the Plateau's Problems, (to appear)

Domain and Boundary Approach in BEM Thermoelasticity

G. Karami

Department of Mechanical Engineering, School of Engineering, Shiraz University, Shiraz, Iran

ABSTRACT

In this paper the application of the BEM to thermoelastic problems is presented. In one approach (domain approach), efforts are not made to transform the integrals due to the temperature to the boundary. Thus, to evaluate these integrals, the domain must be discretized. In another approach (boundary approach) the derivatives of the temperature can be treated as a kind of pseudo-body force. Using this pseudo-body force approach, the integrals due to temperature change and the body force itself can be written in the form of boundary integrals. The formulations are restricted to two-dimensional elastic isotropic martials. Examples are presented to check the accuracy of the formulations, as well as the efficiency of BEM for thermoelastic problems.

INTRODUCTION

The BEM as a powerful numerical tool in engineering analysis is now well established (see for example, Brebbia, Telles and Wrobel [6]). The application of the method to thermoelastic problems were first introduced by Rizzo and Shippy [1]. They used the well-known fundamental solutions of the Navier equations governing displacements in an elastic solid. Wrobel and Brebbia [2] presented boundary elements for steady state heat conduction which can be used in the thermoelastic analysis of solid bodies. Ghosh and Mukherjee [3] developed a boundary element formulation for nonhomogeneous media. An indirect boundary element formulations for thermal analysis is presented by Wearing and Sheikh [4]. The treatment of the thermoelastic problems with either concentrated or distributed heat sources is suggested by Niku and Brebbia [5].

This paper presents both the domain as well as the boundary treatments of the integral equations due to thermal loads for elastic bodies. In the boundary approach the derivative of the temperature is assumed as a kind of bodyforce, therefore using the pseudo-bodyforce approach the domain integrals due to thermal loads can be transformed over the boundary and therefore the domain discretization would not be necessary. The domain approach has the advantage if α the coefficient of expansion is not uniform. In adding the integrals due to bodyforce and thermal loads the basic features of the elastostatic formulations still apply, only, the modifications essentially concern with the evaluation of the known vector, b, in the final system of linear equations [A]z = b, where [A] is the coefficient matrix and z is the unknown nodal vectors.

GOVERNING EQUATIONS FOR 2-D THERMOELASTICITY

Consider a homogenous isotropic solid body subjected to external and internal forces and heating, and furthermore assume that the material is linear elastic and that is stress free at a uniform temperature T_0 when all the external forces are removed. The temperature T_0 is called the reference temperature. Also the assumption made that each component of the strain tensor ϵ_{ij} can be additively decomposed into an elastic and a thermoelastic part ϵ^e_{ij} and ϵ^T_{ij}, respectively. Thus,

$$\epsilon_{ij} = \epsilon^e_{ij} + \epsilon^T_{ij} = \epsilon^e_{ij} + \alpha T \delta_{ij} \tag{1}$$

where α is the coefficient of expansion, T is the instantaneous temperature (the temperature above T_0) and δ_{ij} is the Kronecker delta. The indices take the range 1,2. By noting that $\epsilon_{33}=0$ in plane strain and $\sigma_{33}=0$ in plane stress, the elastic stress strain equations (Hooke's law) may be written as;

$$\sigma_{ij} = 2\mu\epsilon_{ij} + \delta_{ij}\, 2\mu\nu/(1-2\nu)\epsilon^e_{kk} - \delta_{ij} 2\mu(1+\nu)/(1-2\nu)\alpha T \tag{2a}$$

$$\sigma_{ij} = 2\mu\epsilon_{ij} + \delta_{ij} 2\mu\nu/(1-2\nu)\epsilon^e_{kk} - \delta_{ij} 2\mu/(1-2\nu)\alpha T \tag{2b}$$

In the above equations and subsequently, part (a) and (b) of an equation apply to plane strain and plane stress, respectively.

The total strain is related to the displacements u_i through the usual kinematic relations

$$\epsilon_{ij} = \tfrac{1}{2}(u_{i,j} + u_{j,i}) \tag{3}$$

and the equilibrium equations as

$$\sigma_{ij,j} + f_i = 0 \tag{4}$$

On combining the above equations the resulted Navier equation in two dimensions can be written in the form of

$$u_{i,jj} + 1/(1-2\nu)\, u_{j,ji} = -f_i/\mu + 2(1+\nu)/(1-2\nu)\alpha T_{,i} \tag{5a}$$

$$u_{i,jj} + 1/(1-2\nu)\, u_{j,ji} = -f_i/\mu + 2/(1-2\nu)\alpha T_{,i} \tag{5a}$$

In the above equations the effective values of the Poisson's ratio should be used (see Karami and Fenner [8]).

THE BEM FORMULATIONS FOR 2-D THERMOELASTIC PROBLEMS

The displacements at a boundary point P of an elastic body may take one of the two alternative forms, depending on whether the thermal strains are treated as either initial strains or psuedobody-forces. The temperature change and body forces may thus be treated using the two different ways, namely either, (i) the domain approach, or (ii) the boundary approach.

The Domain Approach

Somigliana's boundary identity relating boundary tractions t_i, boundary displacements u_i, body force f_i and the thermoelastic strain ϵ^T_{ij} for a domain S with boundary s may be written as (see for example Lee [11])

$$C_{ki}(P)u_i(P) + \int_s T_{ki}(P,Q)u_i(Q)ds_Q = \int_s U_{ki}(P,Q)t_i(Q)ds_Q$$

$$+ \int_S U_{ki}(P,q)f_i(Q)dS_q + \int_S \Sigma_{kij}(P,q)\epsilon^T_{ij}(p)dS_q \tag{6}$$

in which $P,Q \in s$, $q \in S$; U_{ij} and T_{ij} are the fundamental Kelvin tensors and C_{ij} is a matrix of numbers dependent only upon the local geometry of s at P. The matrices U_{ij}, T_{ij}, Σ_{kij} and C_{ij} for two-dimensional elastostatics are given as

$$U_{ij}(p,Q) = -k_1/\mu\ \{\delta_{ij}(3-4\nu)\ln r - r_{,i}r_{,j}\} \tag{7}$$

$$T_{ij}(p,Q) = -2k_1(1-2\nu)/r\ \{(\delta_{ij} + 2/(1-2\nu)\, r_{,i}r_{,j})\partial r/\partial n$$

$$-r_{,}n_j + r_{,j}n_i\} \tag{8}$$

where,

$k_1 = (1-2\nu)/16\pi(1-\nu)$.

$$\Sigma_{kij}(p,q) = \mu\ \{ U_{ki,j}(p,q) + U_{kj,i}(p,q)\} + \delta_{ij}2\mu\nu/(1-2\nu)$$

$$x U_{kh,h}(p,q) \tag{9a}$$

$$\Sigma_{kij}(p,q) = \mu\ \{ U_{ki,j}(p,q) + U_{kj,i}(p,q)\} + \delta_{ij}2\mu\nu/(1-2\nu)$$

$$x U_{kh,h}(p,q) \tag{9b}$$

$$C_{ij}(P) = \delta_{ij} + \lim_{\varsigma\to 0} \int_{s(\varsigma)} T_{ij}(P,Q)ds_Q \tag{10}$$

In the integral equation (6) the bodyforce and thermal strain integrals are in the form of domain integerals and the thermal strains are assumed as initial strains.

Having the integral identities for the displacements at the interior points the integral equations for the stresses and strains at interior points may be found using the strain-displacement and stress-strain relationships.

The Boundary approach
In the boundary approach, the integrals due to the bodyforces and the thermal strains are converted into boundary integrals. The boundary approach involves the use of the psuedovariables f'_i, t'_i and σ'_{ij}, the definitions of which are

$$f'_i = f_i - \Omega T_{,i} \tag{11}$$

$$t'_i = t_i - \Omega T_{,i} \tag{12}$$

$$\sigma'_{ij} = \sigma_{ij} - \delta_{ij}\Omega T_{,i} \tag{13}$$

where the thermoelastic constant Ω is defined in terms of the shear modulus, effective Poisson's ratio and the coefficient of expansion as follows

$$\Omega = 2\mu\alpha(1+\nu)/(1-2\nu) \tag{14a}$$

$$\Omega = 2\mu\alpha/(1-2\nu) \tag{14b}$$

If the psuedotractions are assumed explicitly as primary variables, the integral contributions due to psuedo-bodyforces by using Green Gauss theorm can be found in the form of boundary integrals (see, Karami [7])

$$I_i(P) = -\int_S U_{ij}(P,Q)\O(Q)n_j(Q)ds_Q + k_1/\mu\int_S \{\partial\O/\partial n\ [2\ln r -1]rr_{,i} - \O(Q)\{(2\ln r-1)n_i(Q) + 2r_{,i}\partial r/\partial n) - k_0 r^2(\ln r-1)n_i(Q)\}ds_Q \tag{15}$$

where $\O$ is the combined scalar potential due to the body and psuedobodyforces and k_0 is a constant to be zero for harmonic functions. $I_i(P)$ is the equivalent to the domain integrals in equation (6).

The use of pseudotractions instead of tractions further complicates the numerical formulations, because the boundary conditions in a physical problem are defined in terms of displacements and tractions. However by separating the potentials due to the actual body force and thermal forces, the use of psuedotractions may be obviated, in which case the additional contribution to b becomes

$$I_i(P) = -\int_S U_{ij}(P,Q)\emptyset^B(Q)n_j(Q)ds_Q + k_1/\mu\int_S\{\partial\emptyset/\partial n$$

$$x[2\ln r-1]rr_{,i} -\emptyset(Q)\{(2\ln r-1)n_i(Q) + 2r_{,i}\partial r/\partial n)$$

$$-k_0r^2(\ln r-1)n_i(Q)\}ds_Q \qquad (16)$$

In equations (15) and (16)

$$\emptyset(Q) = \emptyset^T(Q) + \emptyset^B(Q) \qquad (17)$$

where,

$$\emptyset^T{}_{,i}(Q) = \Omega T_{,i}(Q) \qquad (18)$$

$$\emptyset^B{}_{,i}(Q) = f_{,i}(Q) \qquad (19)$$

Using pseudo-bodyforce approach and by directly applying the total strain-displacement relations and also stress-strain relationship, the integral equations for the strain and stress at interior points may be derived in the form of

$$\epsilon_{ij}(p) = -\int_S S'_{kij}(p,Q)u_k(Q)ds_Q + \int_S D'_{kij}(p,Q)t'_i(Q)ds_Q$$

$$+ \int_S D_{kij}(p,q)f'_k(Q)dS_q \qquad (20)$$

$$\sigma_{ij}(p) = -\int_S S_{kij}(p,Q)u_k(Q)ds_Q + \int_S D_{kij}(p,Q)t'_i(Q)ds_Q$$

$$+ \int_S D_{kij}(p,q)f'_k(Q)dS_q - 2\mu(1+\nu)/(1-2\nu)\ \epsilon^T{}_{ij}(p) \qquad (21a)$$

$$\sigma_{ij}(p) = -\int_S S_{kij}(p,Q)u_k(Q)ds_Q + \int_S D_{kij}(p,Q)t'_i(Q)ds_Q$$

$$+ \int_S D_{kij}(p,q)f'_k(Q)dS_q - 2\mu/(1-2\nu)\ \epsilon^T{}_{ij}(p) \qquad (21b)$$

The matrices S'_{kij}, D'_{kij}, S_{kij} and D_{kij} are 3rd order tensors. The Green Gauss theorem cab be used to transform the integrals over the domain to the boundary. Furthermore the use of pseudo-tractions would complicate the implementation of the boundary conditions. But, by separation of the potentials due to the temperatures and bodyforces, the contributions due to the body force and temperature to the internal strain can be written in the form of boundary integrals as

$$I_{kh}(p) = -\int_S D'_{ikh}(p,Q)\emptyset^B(Q)n_j(Q)ds_Q + k_1/\mu\int_S 2\emptyset/r\ \{\delta_{kh}$$

$$-2r_{,k}r_{,h})\partial r/\partial n + r_{,k}n_h(Q) + r_{,k}n_k(Q)\}$$

$$-\partial\phi/\partial n \ \{\delta_{kh}(2\ln r - 1) + 2r_{,k}r_{,h}\} + k_o/2\{($$

$$2\ln r -1)r\{r_{,k}n_h(Q) + r_{,h}n_k(Q)\}\}ds_Q \qquad (22)$$

Similarly the contribution to the internal stresses can be simplified in terms of boundary integrals and in terms of boundary tractions and displacements.

NUMERICAL FORMULATIONS

The numerical solutions to the integral equations entail the discretization of the boundary and the domain of the body if the domain approach is to be employed. Isoparametric quadratic boundary elements are to be used here, thus the geometry is defined in terms of quadratic shape functions of an intrinsic coordinate Ç which varies from -1 to +1 along the element. Also in such a case, eight-nodded quadrilateral or six-nodded triangular elements can be employed so that, the geometry of the interior elements is governed by quadratic shape functions of two intrinsic coordinates $ç_1$ and $ç_2$ both of which varies from -1 to +1. The shape functions for the geometry of the elements are also used for parametric representation of displacements, tractions, body forces and thermal strains over the elements, that is

$$u_i(Ç) = M^c(Ç)u_i^c \qquad (23)$$

$$t_i(Ç) = M^c(Ç)t_i^c \qquad (24)$$

$$f_i(ç) = N^f(ç)f_i^f \qquad (25)$$

$$\epsilon^T(ç) = M^f(ç)[\epsilon^T_{ij}]^f \qquad (26)$$

$M^c(Ç)$ (c=1,3) is the quadratic shape functions over the line elements and is a function of the intrinsic coordinates Ç. $N^f(ç)$ (for quadrilateral elements, f=1,8 and for triangular elements f=1,6) is the quadratic shape function over the surface elements and is a function of the intrinsic coordinates $ç_1$ and $ç_2$.

By substituting the above parametric representations of the variables and the geometry into the related integrals, the discretized form of the integral equations for the displacements, body forces and thermal strains may be derived in the from of

$$C_{ki}(P^a)u_i(P^a) + \sum_{b=1}^{m} \sum_{c=1}^{3} u_i(P^{d(b,c)}) \int_{S_b} T_{ki}(P^a,Q(Ç))M^c(Ç)$$

$$x J_s(Ç)dÇ = \sum_{b=1}^{m} \sum_{c=1}^{3} t_i(P^{d(b,c)}) \int_{s_b} U_{ki}(P^a,Q(Ç))M^c(Ç)J_s(Ç)dÇ$$

$$+ \sum_{e=1}^{n} \sum_{f}^{6,8} f_i(P^{g(e,f)}) \int_{S_e} U_{ki}(P^a,q(ç))N^f(ç)J_s(ç)dç$$

$$+\sum_{e=1}^{n} \sum_{f=1}^{6,8} \epsilon^T{}_{ki}(P^{g(e,f)}) \int_{S_e} \Sigma_{kij}(P^a,q(ç))N^f(ç)J_s(ç)dç \qquad (27)$$

where P^a is the node at which the integral equation is written, $P^{d(b,c)}$ denotes the cth node of the bth boundary element and $P^{g(e,f)}$ denotes the fth node of the eth interior element. Also, s_b and S_e are the bth boundary element and eth interior element respectively. $J_s(Ç)$ is the jacobian of transformation. m and n are the total number of quadratic boundary and interior elements, respectively.

The set of equations represented by the above equation contains 8m boundary nodal values of displacements and tractions and 3h nodal values of thermal strain where h is the total number of domain nodes required for the description of the interior elements. Clearly h is less than or equal to the sum of the boundary and interior nodes.

Also in the same manner the internal strains and stresses may be evaluated expressed completely as integrals in terms of the nodal variables and the intrinsic coordinates. It is clear that in the domain approach the nodal values of the temperature and the body force need to be specified as part of the data for the problem. This is not a difficult task once you have the temperature and the temperature gradient over the boundary by utilizing the fact that (see for example Brebbia, Telles and Wrobel [6]).

$$\emptyset(p) = 1/4\pi \int_S \{ 2/r\, \partial r/\partial n\, \emptyset(Q) - 2\ln r \partial\emptyset(Q)/\partial n + (2\ln r-1)r$$

$$x\, \partial r/\partial n\, k_0\}ds \qquad (28)$$

With the boundary approach the variation of the potentials and the potential gradients over the boundary can be expressed similar to the displacements and tractions in terms of nodal values; that is

$$\emptyset_i(Ç) = M^c(Ç)\emptyset_i{}^c \qquad (29)$$

$$\partial\emptyset_i/\partial n(Ç) = M^c(Ç)\partial\emptyset_i{}^c/\partial n \qquad (30)$$

Having the parametric representations of the variables in terms of their nodal values, the discretized form of the pseudoboundary integral equations over the boundary can be written similar to the case of the domain approach.

NUMERICAL EXAMPLE

A simple Block under increasing temperature (plane strain)

A block as shown in figure 1 is constrained not to move in y-directions at the top and bottom sides and is under an increasing temperature. The material properties are E=100 Gpa, ν=0.3. For three different cases: (a) a uniform temperature increase T=100 ^{0}K and a uniform $\alpha=10^{-5}$ $^0K^{-1}$, (b) the temperature at the top half T=200 ^{0}K and at the bottom half T=100 ^{0}K and a uniform $\alpha=10^{-5}$ $^0K^{-1}$, (c) a uniform temperature T=100 ^{0}K, and α on the top half=3×10^{-5} $^0K^{-1}$and α on the bottom half =10^{-5} $^0K^{-1}$, the developed stresses and displacements of the central layer are calculated and compared with the exact solution (see for example Goodier [12]) and are presented in table 1. The accuracy is very well satisfied.

Table 1: The BEM results for σ_Y and u_Y

	σ_Y (N/m^2) BEM	σ_Y (N/m^2) Goodier [12]	u_Y(m) BEM
case (a)	-1.42568×10^8	-1.42857×10^8	0
case (b)	-2.14232×10^8	-2.14285×10^8	-6.5×10^{-4}
case (c)	-2.85674×10^8	-2.85714×10^8	-1.3×10^{-3}

An Elastic Slab Between two Rigid Rollers (plane Strain)

An elastic slab (as shown in figure 2) is assumed under an increasing temperature resting between two rigid rollers. The slab is assumed to be made of almunium with E= 68.75 Gpa, ν=0.3 $\alpha=0.3\times10^{-5}$ $^0K^{-1}$. Two different cases of (a) uniform temperature increase and (b) a linear increase in the temperature increase from the surface to the center are assumed. The slab expands under the increase of the temperature and makes contact with the rollers. The contact areas are measured using the BEM for contact problems (see Karami & Fenner [8]). The resulted contact compressive stresses are plotted for both cases in figure 3. Three different curves for three different levels of temperatures for case (a) and three different level of temperatures for case (b) are shown. The temperatures at the center of the slab for case (b) are exactly twice that of equivalent uniform temperature in case (a) to produce the same contact stress. The solutions are compared with the Hertz contact stresses produced under the same load (thermal loads in the contact area due to thermal expansion) in the same figure.

As shown the accuracy is very well satisfied. In this example the boundary approach was employed. The boundary discretization of a quarter of the bodies is shown in figure 2.

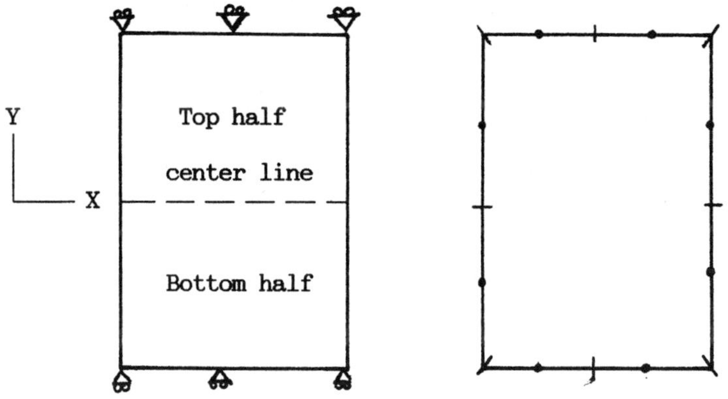

(a) (b)

Fig. 1. An elastic block under the increasing temperature
(a): The geometry
(b): Boundary discretization

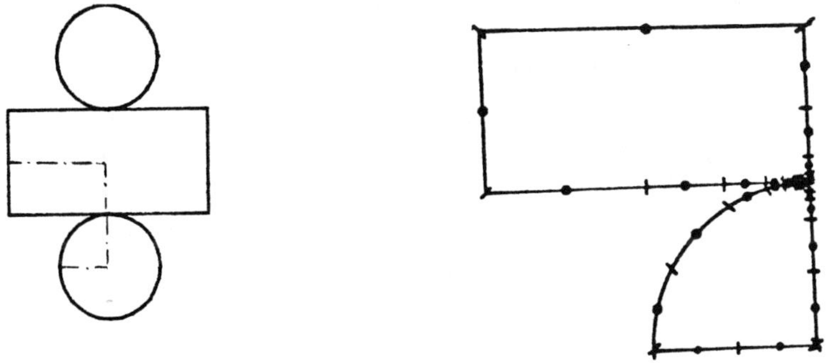

(a) (b)

Radius of rigid roller = 1 m

Thickness of slab = 2 m

Fig. 2. An elastic slab between two rigid rollers
(a): The geometry
(b): Boundary discretization

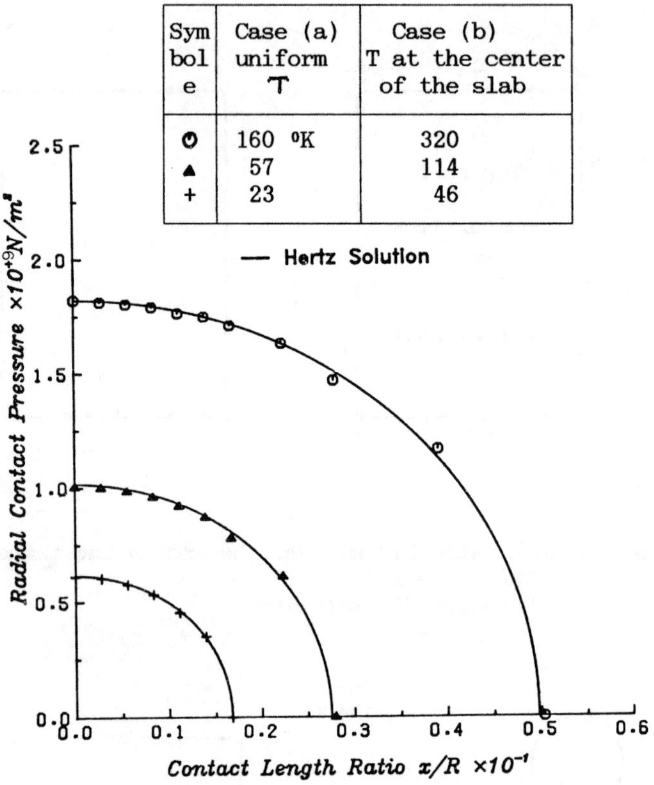

Fig. 3. The compressive stresses between the slab and the rigid roller due to increase in temperature of the slab

CONCLUSION

A theoretical and an isoparametrical quadratic boundary element formulation for two-dimensional thermoelastic problems was presented. The psuedobody force approach is simpler and easier to be employed in comparison with the domain approach. However, for nonuniform α the boundary approach doesn't work.

REFERENCES

1. Rizzo, F.J. and Shippy, D.J. - An advanced boundary integral equation method for three-dimensional thermoelasticity, Int. Journal for Num. Meth. in Eng., Vol. 11,pp. 1753-1768, 1977.
2. Wrobel L.C. and Brebbia, C.A.- Boundary elements in thermal problems in Numerical Methods in Heat Transfer, (Ed., R.W. Lewis), John Wiley, Chichester, 1981.
3. Ghosh, S. and Mukherjee, S. Boundary element method analysis of thermoelastic deformation in nonhomogeneous media, Int. J. Solids Structures, Vol. 20,pp. 829-843 1984.
4. Wearing J.L. and Sheikh, M.A. A regular indirect boundary element method for thermal analysis, Int. Journal for Num. Meth. in Eng., Vol. 25,pp. 495-515, 1988.
5. Niku, S.M. and Brebbia C.A. The treatment of thermoelastic problems with concentrated and distributed heat sources, in Boundary Element X, Vol. 3. Stress Analysis, Springer-Verlag, Berlin and New York, 1988.
6. Brebbia, C.A., Telles, J.S.F. and Wrobel, L.C., Boundary Element Method Techniques -Theory and Applications in Engineering, Springer-Verlag, Berlin and New york, 1988.
7. Karami, G., Boundary element method for elastic thermoelastic bodyforce problems using pseudo-body force approach, in Computational Mechanics '88 Theory and Applications (Eds. S.N. Atluri, and G. Yagawa), Vol 1, Springer-Verlag, Berlin, 1988.
8. Karami, G. and Fenner, R.T. - A Two-Dimensional BEM for Thermo-Elastic Body Forces Contact Problems. In Boundary Element IX, vol. 2: Stress Analysis Applications (eds. C.A. Brebbia, W.L. Wendland, G. Kuhn), Springer- Verlag, Berlin, pp. 417-437,1987.
9. Karami, G. and Fenner, R.T. - Application of boundary integral equation method (BIE)to two-dimensional elastic contact problem using isoparametric quadratic elements. Iranian J. Sci. Tech., Vol. 11, pp. 153-176, 1987.
11. Lee, K.H., A Boundary Integral Equation Method for Two-dimensional Elastoplastic Analysis. Ph.d Thesis, Univ. of London, 1983.
12. Goodier, J.N., Thermal Stress, Trans. ASME, J. of Appl. Mech., Vol. 4, A33, 1937.

A General Boundary Integral Formulation for the Numerical Solution of Bending Multilayer Sandwich Plates

G. Davi
Istituto di Aeronautica, Facolta di Ingegneria, Universita di Palermo, Palermo, Italy

ABSTRACT

A direct approach is employed to obtain a general formulation for the analysis of multilayer sandwich plates in terms of integral equations, involving displacements, rotations, moments and shear on the plate boundary. These equations are coupled with prescribed boundary conditions, involving the same variables and give a convenient basis for numerical solution. A simple discretization scheme is described and two examples are treated to show the nature and quality of some typical results.

INTRODUCTION

The integral equation method [1-2] is used for solution of thin plate bending problems [3-5], taking Kirchhoff's theory as valid, i. e. assuming shear deformation to be negligible. Other authors [6-8] have suggested a more general formulation of bending plate problems, using Reissner's plate theory [9], which accounts for the transverse bending shear deformation and retains the three natural boundary conditions.

In this paper the method is extended to the structural analysis of bending multilayer isotropic sandwich plates. Among the various theories proposed for the formulation of the problem of sandwich plates, that reported in [11-12] is adopted. In this formulation the governing equations for bending of multilayered sandwich plates are developed by variational methods. These equations are analogous to those obtained by Mindlin [10] for the generalized bending plate theory.

EQUATIONS OF EQUILIBRIUM AND COMPATIBILITY

A multilayer sandwich plate is considered (fig. 1), consisting of n-1 cores of thicknesses $h_1, h_2, ...h_{n-1}$ and n facing membranes of thicknesses $t_1, t_2, ...t_n$. The plate is referred to an $x_1 x_2 x_3$ orthogonal system. The $x_1 x_2$ plane is assumed to be parallel to the undeformed surface of the plate. Let $z_1, z_2, ...z_n$ be the distances measured from the $x_1 x_2$ plane to the middle plane of each membrane. Each facing and each core is assumed to be isotropic and homogeneous and to possess different elastic properties.

The following assumptions are essential for this analysis:
- the transverse rigidities of the core are relatively high compared to the facings, so that the shear is carried by the core, which in turn are assumed to be devoid of flexural rigidity.
- the total thickness is still small in comparison with the dimensions in other directions.

Let σ_{ij} $(i,j = 1,2)$ be the stress in the generic membrane, and σ_{i3} the shearing stress in the generic core. Then the moments and shear forces are given by the following relations:

$$\begin{aligned} M_{ij} &= \Sigma\, z_l t_l \sigma_{ij} \\ T_{i3} &= \Sigma\, h_s\, \sigma_{i3} \\ N_{ij} &= \Sigma\, t_l \sigma_{ij} = 0 \end{aligned} \tag{1}$$

Under these assumptions the equilibrium equations become:

$$\underline{D}\,\underline{S} = -\underline{f} \tag{2}$$

where:

$$\underline{D} = \begin{bmatrix} \partial/\partial x_1 & 0 & \partial/\partial x_2 & -1 & 0 \\ 0 & \partial/\partial x_2 & \partial/\partial x_1 & 0 & -1 \\ 0 & 0 & 0 & \partial/\partial x_1 & \partial/\partial x_2 \end{bmatrix}$$

$$\underline{S}^T = \left[M_{11}, M_{22}, M_{12}, T_{13}, T_{23}\right] \tag{3}$$

$$\underline{f}^T = \left[m_1, \ m_2, \ q\right] \tag{4}$$

In equation (4) m_1 and m_2 are the external moments applied per unit area, q is the transverse load per unit area applied normal to the plate. Let ϑ_1 and ϑ_2 be the rotations of generic segment orthogonal to the reference plane $x_1 x_2$ and let w be the transverse displacement. Then the generalized strain components are given by:

$$\underline{\varepsilon} = \underline{C}\ \underline{u} \tag{5}$$

where:

$$\underline{\varepsilon} = \begin{bmatrix} \varepsilon_{11} \\ \varepsilon_{22} \\ \varepsilon_{12} \\ \gamma_{13} \\ \gamma_{23} \end{bmatrix} \ ; \quad \underline{C} = \begin{bmatrix} \partial/\partial x_1 & 0 & 0 \\ 0 & \partial/\partial x_2 & 0 \\ \partial/\partial x_2 & \partial/\partial x_1 & 0 \\ 1 & 0 & \partial/\partial x_1 \\ 0 & 1 & \partial/\partial x_2 \end{bmatrix} \ ; \quad \underline{u} = \begin{bmatrix} \vartheta_1 \\ \vartheta_2 \\ w \end{bmatrix}$$

The moments and transverse shear forces, as a function of generalized displacements, are given by:

$$\underline{S} = \underline{E}\ \underline{\varepsilon} = \underline{E}\ \underline{C}\ \underline{u} \tag{6}$$

where $\underline{E}$ is the rigidity matrix defined (see references [11-12]) as follows:

$$\underline{E} = d \begin{bmatrix} 1 & \nu & 0 & 0 & 0 \\ \nu & 1 & 0 & 0 & 0 \\ 0 & 0 & (1-\nu)/2 & 0 & 0 \\ 0 & 0 & 0 & g/d & 0 \\ 0 & 0 & 0 & 0 & g/d \end{bmatrix} \tag{7}$$

given that d is flexural rigidity, ν the equivalent Poisson's ratio and g shear rigidity:

$$\begin{aligned} d &= \Sigma d_l = \Sigma\ E_l t_l z_l ((z_l - z_o) - (1+\nu_l)\zeta)/(1-\nu_l^2) \\ \nu &= \Sigma \nu_l d_l / d \\ g &= \Sigma\ h_s G_s \end{aligned} \tag{8}$$

In equation (8) E_l and ν_l represent the elastic constants of the generic membrane, G_s is the shear modulus of the generic core. The following relations are also put:

$$z_o = (\Sigma\ E_l t_l z/(1+\nu_l))/(\Sigma\ E_l t_l/(1+\nu_l))$$
$$\zeta = (\Sigma\ E_l t_l (z-z_o)/(1-\nu_l))/(2\ \Sigma\ E_l t_l/(1-\nu_l)) \tag{9}$$

In the case when ζ is equal to zero, z_o represents the ordinate of the neutral surface.

INTEGRAL EQUATIONS

Let us consider the plate, however constrained on the boundary Γ, and subjected in the region Ω to a set of forces, moments and transverse loads, however distributed. Let $\underline{u}$ be the generalized displacements associated with the external forces $\underline{f}$ and with the tractions $\underline{t}$ acting on the boundary. Let $\underline{u}_j$ and $\underline{t}_j$ be the generalized displacements and the tractions due to a system of fictitious external forces $\underline{f}_j$. Applying the theorem of reciprocity to the two systems, we have:

$$\int_\Gamma (\underline{u}_j{}^T \underline{t} - \underline{t}_j{}^T \underline{u})\ d\Gamma = \int_\Omega (\underline{f}_j{}^T \underline{u} - \underline{u}_j{}^T \underline{f})\ d\Omega \tag{10}$$

Equation (10) is the fundamental relation for the solution of the problem by the boundary element method. Let the system of fictitious forces be so characterized that:

$$f_{ij} = \Delta(p-p_o)\ c_{ij}(p_o) \tag{11}$$

where Δ is the Dirac function and the c_{ij} (p_o) are the components of the unit vector applied at p_o in the j-th direction. Substituting equation (11) into equation (10) we have:

$$\underline{c}_j{}^T \underline{u}(p_o) = \int_\Gamma (\underline{u}_j{}^T \underline{t} - \underline{t}_j{}^T \underline{u})\ d\Gamma + \int_\Omega \underline{u}_j{}^T \underline{f}\ d\Omega \tag{12}$$

For p_o on the boundary, and for $j = 1,3$, equation (12) provides the boundary integral equation, which bind between them the unknown boundary data. Putting:

$$\underline{c}^* = [c_{ij}]^T;\quad \underline{u}^* = [u_{ij}]^T;\quad \underline{t}^* = [\underline{t}_{ij}]^T$$

the integral equations can be written:

$$\underline{c}^* \underline{u}(p_o) = \int_\Gamma (\underline{u}^* \underline{t} - \underline{t}^* \underline{u})\, d\Gamma + \int_\Omega \underline{u}^* \underline{f}\, d\Omega \qquad (13)$$

Equation (13), together with prescribed boundary data, should be sufficient to determine remaining boundary data needed in equation (12) to characterize the solution of the original problem.

FUNDAMENTAL SOLUTIONS

The fundamental solutions associated with the system of forces $\underline{f}_j$, are particular integrals of the equilibrium equations (2) which, expressed as a function of the generalized displacements, become :

$$\underline{D}\ \underline{E}\ \underline{C}\ \underline{u}_j = -\ \underline{f}_j \qquad (14)$$

Integrating equation (14), and taking into account equation (11), we have :

$$\begin{aligned} u_{ij} &= -\frac{\partial}{\partial x_i}(\bar{x}_j \varphi/2\ d + \frac{\partial}{\partial x_j}(\varphi - \psi)/g) - \delta_{ij}\lambda^2\psi/g \\ u_{3j} &= -\ u_{j3} = \bar{x}_j \varphi/2\ d \qquad i,j = 1,2 \qquad (15) \\ u_{33} &= (r^2(4\ \pi\ \varphi - 1)/4\ \pi\ d - 4\ \varphi/g)/4 \end{aligned}$$

where :

$$\lambda^2 = 2\ g/(1 - \nu)d$$

$$\bar{x}_j = x_j(p) - x_j(p_o)$$

$$\varphi = (\log\ r)/2\ \pi$$

$$\psi = (K_o(\lambda r))/2\ \pi\ \lambda^2$$

where $r(p,p_o)$ is the distance of point p from the point p_o and $K_o(\lambda r)$ is the modified Bessel function of the second kind of order zero.

The boundary tractions are given by:

$$\underline{t}_j = \underline{D}n\ \underline{S}_j \qquad (16)$$

where we put:

$$\underline{D}n = \begin{bmatrix} \alpha_1 & 0 & \alpha_2 & 0 & 0 \\ 0 & \alpha_2 & \alpha_1 & 0 & 0 \\ 0 & 0 & 0 & \alpha_1 & \alpha_2 \end{bmatrix}$$

$$\underline{S}_j = \underline{E}\ \underline{C}\ \underline{u}_j \qquad (17)$$

where α_1 and α_2 are direction cosines of the outer normal on the boundary point. The coefficients c_{ij} are obtained from equation (11):

$$\underline{c}_j(p_o) = \int_\Omega \underline{f}_j\, d\Omega = -\int_\Omega \underline{D}\ \underline{E}\ \underline{C}\ \underline{u}_j\, d\Omega \qquad (18)$$

For $p_o \in \Omega, c_{ij} = \delta_{ij}$, while for $p_o \in \Gamma$, if the boundary is smooth at p_o, $c_{ij} = \delta_{ij}/2$.

NUMERICAL SOLUTION

The integral equations (13) can be expressed in numerical form discretizing the boundary into a finite number of elements. When the boundary is discretized, the unknown boundary data are to be defined in terms of their nodal values. Supposing that, for simplicity's sake, the tractions are applied to a part Γ_1 of the boundary, while the displacements are known on the part Γ_2, we have:

$$\underline{u} = \underline{N}\ \underline{\delta}\ ; \quad \underline{t} = \underline{N}\ \underline{\bar{t}} \qquad (19)$$

where $\underline{\delta}$ e $\underline{\bar{t}}$ are the nodal parameters and $\underline{N}$ is the shape function. Substituting equation (19) into equation (13) we have:

$$\underline{c}^*\underline{u}(p_o) = \int_{\Gamma_1} \underline{u}^*\underline{t}\, d\Gamma - \int_{\Gamma_2} \underline{t}^*\underline{u}\, d\Gamma + \int_\Omega \underline{u}^*\underline{f}\, d\Omega +$$

$$+ \int_{\Gamma_2} \underline{u}^*\underline{N}\ \underline{\bar{t}}\, d\Gamma - \int_{\Gamma_1} \underline{t}^*\underline{N}\ \underline{\delta}\, d\Gamma \qquad (20)$$

If the boundary is not regular at the point p_o the coefficients c_{ij} are calculated by trasforming the domain integral that appears in equation (18) into a boundary integral. Taking into account equations (16) and (17), we have:

$$c_{ij}(p_o) = -\int_{\Gamma}(t_{ij} - x_i\, t_{3j})\, d\Gamma \qquad i = 1,2$$

$$c_{3j}(p_o) = -\int_{\Gamma} t_{3j}\, d\Gamma \qquad (21)$$

By locating the point p_o on each boundary node in turn, a linear system of algebraic equations, 3 per node, is obtained. The unknowns are either the nodal generalized displacements or nodal traction components.

In the following applications the boundary is discretized into linear elements, with linear interpolation of the boundary data. The integrals are calculated numerically using Gaussian quadrature.The functions to be integrated present singularities of the type lg $r(p,p_o)$ and $1/r(p,p_o)$. In the first case numerical integration presents no difficulty. In the second case the integral is to be understood as a Cauchy principal value. The numerical integration is obtained by removing the singular part, considering that in general we have:

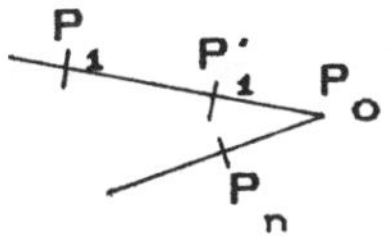

$$\overline{P_oP'_1} = \overline{P_oP_n}$$

$$-\int_{P_n}^{P_o}(1/r)dr + \int_{P_o}^{P_1}(1/r)dr =$$

$$-\int_{P_n}^{P_o}(\ \ )dr + \int_{P_o}^{P'_1}(\ \ )dr + \int_{P'_1}^{P_1}(\ \ )dr = +\int_{P'_1}^{P_1}(1/r)dr \qquad (22)$$

APPLICATIONS

The numerical applications shown concern a square sandwich plate (side length L),assumed to be clamped and simply supported on all four edges, subjected to a uniform load q. The examples refer to two different values of the relationship between shear rigidity and flexural rigidity.

It is taken into account that in the corners the moments and shear forces are zero for the clamped plate, while for the hinged plate only the twisting moments are not zero.

Each edge was discretized into 2, 4, or 8 elements, the results obtained beginning to be appreciable from the second discretization.

Fig.2 demonstrates the convergence of the maximum deflection of the plate when the number of elements increases.Shear results for clamped and simply supported plate are shown in fig. 3. Bending moment results for clamped plate are shown in fig. 4.

The results obtained are in good agreement with those reported in the literature [13-14].

REFERENCES

[1] Rizzo, F.J. An integral equation approach to boundary value problems of classical elastostatic, Quart. Appl. Math., Vol.25, pp.83-95,1967.

[2] Cruse, T.A. An improved boundary-integral equation method for three dimensional elastic stress analysis, Comp. and Struct., vol.4, pp.741-754, 1974.

[3] Bezine, G. and Gamby, D.A. A new integral equation formulation for plate bending problem (Ed. Brebbia, C.A.), pp. 327-342, Recent Advances in Boundary Element Methods. Pentech Press,London, 1978.

[4] Stern, M. A general boundary integral formulation for the numerical solution of plate bending problems, Int. J. Solids Struct., vol. 15, pp. 769-782, 1979

[5] Costa Jr., J.A. and Brebbia, C.A. Plate bending problems using B.E.M. (Ed. Brebbia),pp 3-43 to 3-63, Proc. 6th. Int. Conf. BEM, Southampton, Computational Mechanics Publications, 1984.

[6] Van der Weeën, F. Application of the direct boundary element method to Reissner'r plate model (Ed. Brebbia,C.A.), pp.487-499, Proc.4th. Int. Conf. Boundary Element Method in Engineering,Southampton,Computational Mechanics Centre, 1982.

[7] Van der Weeën, F. Application of the boundary integral equation method to Reissner'r plate model, Int. J. Numer. Engng., Vol. 18, pp.1-10, 1982.

[8] Brebbia, C.A. and Long, S.Y. Boundary Element Analysis of Plates Using Reissner'r Theory (Ed. Brebbia, C.A., Wendlan, W.L. and Kuhn, G.)

Vol.2, pp. 3–18, Proc. 9th. Int. Conf .Boundary Elements IX-Vol.2-Stress Analysis Applications, Computational Mechanics Publications, 1987.

[9] Reissner, E. The Effect of Transverse Shear Deformation on the Bending of Elastic Plates, J. Appl.Mech., vol.12, pp.A69-A77, 1945.

[10] Mindlin, R.D. Influence of rotatory inertia and shear on flexural motions of isotropic,elastic Plates, J. Appl. Mech., vol.18, pp.31-38, 1951.

[11] Liaw, B.D. and Little, R.W. Theory of bending multilayer sandwich plates, AIAA Journal,vol.5, pp. 301-304, 1967.

[12] Azar, J.J. Bending theory for multilayer orthotropic sandwich plates, AIAA Journal, vol. 6, pp. 2166-2169, 1968.

[13] Plantema, F.J. Sandwich Construction, Wiley, New York, 1966.

[14] Tessler, A. On a conforming Mindlin-type plate element (Ed. Whiteman, J.R.), pp. 119-126, The mathematics of finite elements and applications IV, Mafelap, 1981, Academic Press, London, 1982.

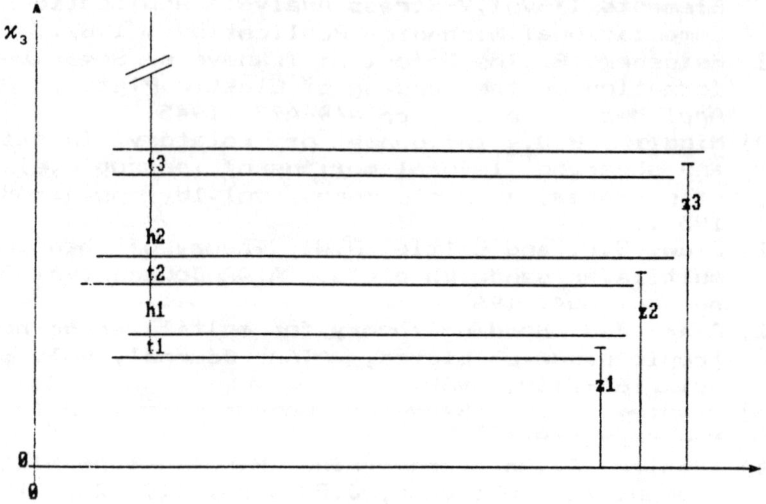

fig. 1 - A cross section of the plate

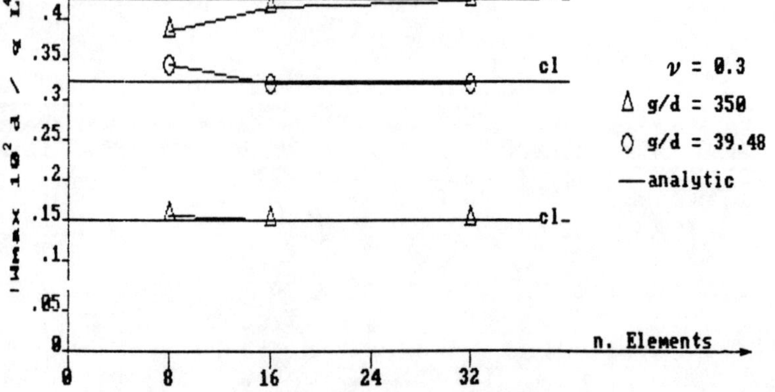

fig. 2 - Convergence of Wmax

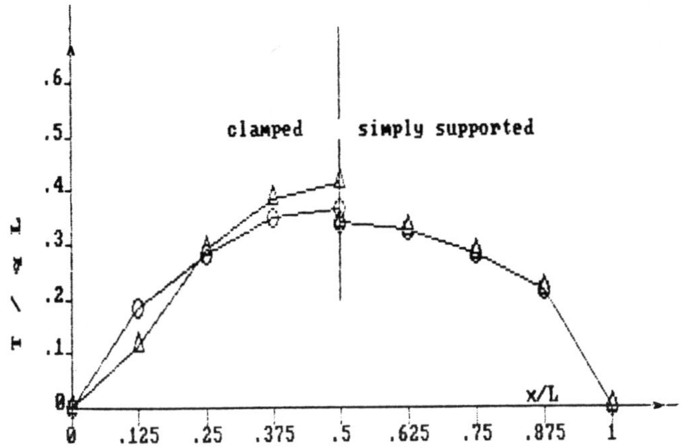

fig. 3- Shear on half an edge

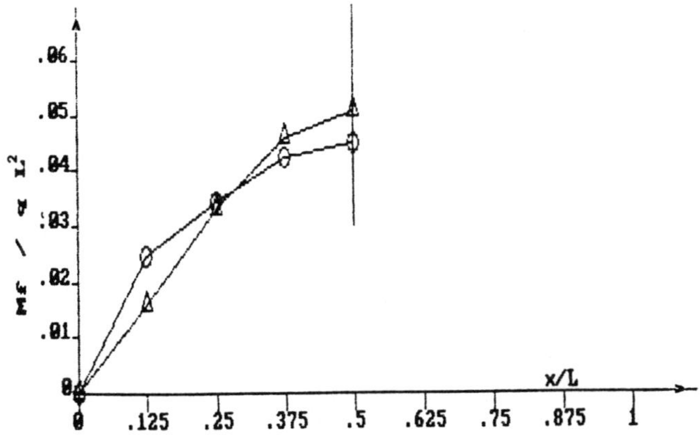

fig. 4 - Bending moment on half an edge

Green's Functions for Inhomogeneous Media for Boundary Elements

Gautam Dasgupta

Department of Civil Engineering and Engineering Mechanics, Columbia University, New York, NY 10027, USA

ABSTRACT

A method is presented to generate **Green's Functions for inhomogeneous anisotropic media** from their counterpart that pertains to the homogeneous unbounded domain. The intent is to immediately employ, with very minor recoding any suitable computer package at hand, in order to construct the inhomogeneous boundary elements. The **principle of flux-invariance** constitutes the kernel of the presentation. The technique also guides to fine-tune by numerically *renormalizing* the finite element shape functions so as to accommodate the material inhomogeneity. For *stochastic* boundary elements and finite elements the proposed formulation can be utilized in conjunction with Monte Carlo simulations.

INTRODUCTION

There are two aspects to the problem solving techniques in the field of computational mechanics. First, the theoretical formulation is carried out in the long-hand symbolic form. Subsequently, the numerical simulation is undertaken in a digital computer. There are instances where a formal solution is possible for a class of problems only within the confines of the numerical modeling. Then it suffices to furnish a development scheme so that an established computer code can be used directly for the modified enlarged class. This paper focuses on such a theoretical formulation. Numerical simulations with the boundary element method, refer to [1], for deterministically or stochastically inhomogeneous and anisotropic media, as illustrated in [2], constitute the wider class where the focus lies on the popular homogeneous isotropic computation.

An extremely strong point for the boundary element method, over its competitors like the finite difference or the finite element method, is that it allows one to examine the problem in its entirety. For engineering problems, which arise out of the elliptic partial differential equation of mathematical physics [3], the strong analyticity, i.e. the infinite continuity property of the solutions, is not compromised until a decision has been made in terms of retaining the order of smoothness in the truncated approximation. This improves tremendously the behavior of the solution in the entire domain. An analogous higher order polynomial-based finite element scheme would introduce non-corner nodes with hierarchic shape functions. The associated "worst case solutions " —in the words of the *information-based complexity theory*— which are signified by the individual "trial functions, " yield unrealistic signatures which do not conform with the mean value property of test functions for elliptic partial differential equations. Controversial adjustments for the evaluation of the energy type integrals, according to the selective and reduced quadrature, and corrective measures for the "mechanism modes " become essential besides tolerating the "overlap and opening " of boundary interfaces [4]. This distinction, of the former Green function-based method, has surfaced as being "more forgiving " than the eigenfunction-based latter one, in terms of handling the error due to spatial discretization. Needless to state that the boundary element procedure uniformly converges for any arbitrary order and spatial dimensionality of elliptic problems described in the references of mathematical physics [3].

Development of a boundary element computer program, at least for the instructional purposes, is relatively simple. Refer to [1]. For commercial usage, computer programs, such as BEASY [5], are available for a variety of personal computers, work stations and mainframes.

The non-sparse compact nature of the input-output transformation matrices makes the procedure extremely amenable to vector processors and for computers with parallel and pipeline architecture. The method does not capitalize in its solution strategy any form of symmetric matrix operation. Hence an implementation for a non–self-adjoint problem is of no consequence. This is indeed an advantage of the method over the conventional finite element one where the symmetry of the force-response transformation matrix indirectly justifies a banded equation of very large length. Of course the arbitrariness of the boundary domain is where the approximation enters in the form of spatial discretization. This has one-to-one—onto association with the degree of smoothness of the overall numerical solution.

However, one shortcoming discouraged a number of users to employ such numerically powerful methods, which are based on the Green function formulation. In almost all cases the material property of the medium to be discretized was required to be homogeneous. Such an obstacle for the application of the method is removed herein. In the generalized formulation and application of the boundary element method, it is crucial to obtain a good set of approximation of the singular Green's function solutions for the field operators. For a large number of partial differential equations of mathematical physics, closed form and very accurate approximate (series type) solutions are available. However, a majority of those field problems are restricted to the *constancy* of the constitutive properties.

The aforementioned Green's function solutions pertain to unbounded media. Hence the *Sommerfeld decay characterization* is used instead of the natural and kinematic boundary conditions. The boundedness, rather than the zero value, of the displacement field *at infinity* with a finite integral for the traction on the infinite contour is already there in the Green's function for the homogeneous isotropic case. Thus the *equivalence* of two partial differential equations with the Dirac's delta on the right hand side *is sufficient*. This observation has led to the **principle of the flux invariance** which has been employed herein.

Particular examples involving the Laplacian are described in this paper. Theoretical development for the Green's functions associated with elliptic problems which have inhomogeneous constitutive properties is sketched out. For self-adjoint cases a numerical scheme is presented to construct the *inhomogeneous Green's functions* from the numerical one that pertains to the homogeneous continuum. The scope of this conference paper does not permit an elaboration with several numerical examples. A journal paper is being prepared to illustrate the **principle of the flux invariance** .

THE FUNDAMENTAL PROBLEM

Field Equations

A generic location, in the n-dimensional spatial domain $\mathcal{D}$, is represented by the set of variables $\{x_i\}$, $i = 1, 2, 3, \ldots, n$. The governing field equation is:

$$\mathcal{L}[u(x_i)] = f(x_i) \tag{1}$$

The forcing function is $f(x_i)$ and the solution for the appropriate boundary value problem is indicated by $u(x_i)$. The relation is described in terms of the **linear** (ordinary *or* partial) differential operator $\mathcal{L}[\bullet]$. The equation is valid in the **open** domain $\mathcal{D} \ni \{x_i\}$. On its boundary $\mathcal{B}$ a set of consistent Dirichlet, Neumann, Robin [3] or the Almansi [6] type condition is prescribed. Symbolically, the boundary data is indicated by:

$$\mathcal{M}[u(x_i)] = h(x_i) \tag{2}$$

The *fundamental* Green's function *solution* $\mathcal{G}$ satisfies the Sommerfeld regularity condition, which is usually written in terms of complex variables even for real-valued problems. This boundary data for the Green's function is invariant for all distribution of the material properties. This is a significant advantage of the Green's function construction for inhomogeneous media where the boundary condition of the partial differential equation does not play any quantitative role. For the linear operator $\mathcal{L}$ of Eq. (1) the *singular* partial differential equation is defined for the Green's function $\mathcal{G}$:

$$\mathcal{L}\{(x_i)\}[\mathcal{G}(x_i|x_i^0)] = \delta(x_i - x_i^0), \tag{3}$$

which is abbreviated to:

$$\mathcal{L}\mathcal{G} = \delta. \tag{4}$$

Therein δ is the Dirac's delta function. This is called the **fundamental problem**, refer to [3] for details. This singular form with the regularity condition in Eq. (2) is of enormous analytical and computational significance for all spatial problems of mathematical physics.

The convolution operator, $\otimes$, is employed to generate a particular solution u of Eq.(1) from $\mathcal{G}$ of Eq.(3) for a prescribed *integrable* forcing function f leading to:

$$u = \mathcal{G} \otimes f. \tag{5}$$

The associated computation for the field variable u involves the following integration over the domain $\mathcal{D}$:

$$u(x_i) = \int_{\mathcal{D}} \mathcal{G}(x_i|x_i^0)f(x_i^0)d\mathcal{V}, \tag{6}$$

where $\mathcal{V}$ is a volume element of $\mathcal{D} \ni \{x_i, x_i^0\}$.

It is important to note that the Green's function is of course one of the many *particular solutions* on the entire n-dimensional space $\mathcal{R}^n \supseteq \mathcal{D}$. The aforementioned equation is indeed valid for any spatial domain of arbitrary dimension n.

Operator Decomposition

In order to derive the *essential* and the *natural* boundary conditions for the governing partial differential equation it is customary to *split* the operator $\mathcal{L}$ into two linear components in the form:

$$\mathcal{L}[\bullet] \doteq \mathcal{L}^{(1)}[\bullet]\mathcal{L}^{(2)}[\bullet]. \tag{7}$$

In the finite element method the essential boundary conditions are first approximated and the natural ones are represented in those terms. It is, in principle, possible to construct Green's functions for the *decomposed* operators such that:

$$\mathcal{L}^{(1)}[\mathcal{G}^{(1)}] = \delta \quad and \quad \mathcal{L}^{(2)}\mathcal{G}^{(2)} = \delta. \tag{8}$$

This is to state that the equilibrium equation is first solved in terms of the flux-type variables and then the lower order derivatives are recovered. The first of Eq. (8) depicts the situation. The flexibility approach of the Green's function then allows one to construct the flux variables irrespective of the constitutive properties. It is then possible to obtain a useful relation between $\mathcal{G}^{(1)}$, $\mathcal{G}^{(2)}$, and $\mathcal{G}$ as follows.

$$\mathcal{L}[u] = f, \quad \Rightarrow \quad \mathcal{L}^{(1)}\left[\mathcal{L}^{(2)}[u]\right] = f, \tag{9}$$

in the decomposed form. Hence

$$\mathcal{L}^{(2)}[u] = \mathcal{G}^{(1)} \otimes f, \quad \text{finally} \quad u = \mathcal{G}^{(2)} \otimes \mathcal{G}^{(1)} \otimes f, \tag{10}$$

leads to:

$$\mathcal{G} = \mathcal{G}^{(2)} \otimes \mathcal{G}^{(1)}. \tag{11}$$

It is convenient, from the *numerical* view point, to represent the aforementioned convolution in the Fourier Transform domain. In there the convolution product gets converted into an ordinary multiplication. Let $\mathcal{F}$ denote the transform operator, then:

$$\mathcal{F}[\mathcal{G}] = \mathcal{F}\left[\mathcal{G}^{(2)}\right] \cdot \mathcal{F}\left[\mathcal{G}^{(1)}\right] \tag{12}$$

Self-adjoint Problems

The boundary element method can be employed for all problems of mathematical physics in an arbitrarily shaped spatial domain $\mathcal{D}$. The *essential* and the *natural* boundary conditions in conjugation define an "energy like " inner-product. In a canonical form it is always possible for those self-adjoint problems to yield the *symmetric decomposition* for the field operator. The *finite element* method for those cases lead to symmetric system (stiffness type) matrices. Then from Eq.(7)

$$\mathcal{L}^{(1)} = \mathcal{L}^{(2)} = \mathcal{L}^{(0)}(\text{say}) \quad \Rightarrow \quad \mathcal{G}^{(1)} = \mathcal{G}^{(2)} = \mathcal{G}^{(0)}(\text{say}). \tag{13}$$

Hence

$$\mathcal{G} = \mathcal{G}^{(0)} \otimes \mathcal{G}^{(0)} \quad \text{or} \quad \mathcal{F}[\mathcal{G}] = \left(\mathcal{F}\left[\mathcal{G}^{(0)}\right]\right)^2. \tag{14}$$

Suppose only the numerical values of $\mathcal{G}$ are given at discrete points. It is then possible to obtain those for the decomposed operator $\mathcal{G}^{(0)}$ at the same set of points.

Higher Order ODES

In a variety of engineering applications one encounters the following class of *one-dimensional* problems for which the boundary element method is frequently sought. In terms of mathematical physics these problems are termed to be the ordinary differential equations *with* **quasi-derivatives.** Such an equation of order $2p$ is written in the form:

$$u^{[2p]}(x) = f(x). \tag{15}$$

The *quasi-derivatives*, $u^{[k]}$, for $k \leq 2p$ are indicated with superscripts enclosed in square brackets, [and]. In this paper, in the interest of clear distinction, the k th. ordinary differential coefficient will be indicated with superscripts encased between parentheses, (and):

$$u^{(k)} = \frac{d^k u}{dx^k}. \tag{16}$$

These *quasi-derivatives* are identical with the ordinary ones for the *essential* (*geometrical* or the *kinematic*) boundary conditions. Hence:

$$u^{[0]} = u, \quad u^{[1]} = \frac{du}{dx}, \ldots u^{[k]} = \frac{d^k u}{dx^k}, \quad \text{for} \quad 0 \leq k \leq (p-1). \tag{17}$$

On the other hand, in the expressions for the *natural* or the *force* boundary conditions "constitutive like " variables, $c_1(x), c_2(x), \ldots, c_p(x)$, appear to be **functions of** x , not merely constants, as in the homogeneous field problems. Hence the *higher quasi-derivatives* are defined to be:

$$u^{[p]} = c_r \frac{d^p u}{dx^p}, \quad u^{[p+1]} = \frac{du^{[p]}}{dx} + c_{r-1} u^{[p-1]},$$
$$u^{[p+2]} = \frac{du^{[p+1]}}{dx} + c_{r-2} u^{[p-2]}, \ldots, \tag{18}$$

The motivation is, obviously, to capture the **correct form** of the natural boundary conditions for the inhomogeneous constitutive properties. This generic problem corresponds to the case when the operators in Eqs.(7) and (13) lead to:

$$\{\mathcal{L}^0\} = \left\{ \frac{d^p}{dx^p}, \frac{d^{p-1}}{dx^{p-1}}, \frac{d^{p-2}}{dx^{p-2}}, \ldots, \frac{d}{dx} \right\} u(x). \tag{19}$$

The constitutive tensor is a diagonal matrix encased between $\lceil and \rfloor$:

$$[C] = \lceil c_p(x), c_{p-1}(x), \ldots, c_1(x), c_0(x) \rfloor. \tag{20}$$

In order to ensure stability, for each individual diagonal coefficient, $(-1)^k c_k(x)$, should be a non-zero positive function of x. It is obvious that:

$$u^{[2p]} = \left(c_p(x) u^{(p)}\right)^{(p)} + \left(c_{p-1}(x) u^{(p-1)}\right)^{(p-1)} + \ldots + (c_1 u^{(1)})^{(1)} + c_0 u \tag{21}$$

This class of problems can be solved by the technique presented herein according to Eq. (14).

Second Order PDES

The next class of problems is the "vector version " of the aforementioned one with *quasi-derivatives* as demonstrated in Eq.(17). Now there are n independent spatial coordinates, $x_1, x_2, \ldots, x_n$. The partial differential equation is of second order. Here, one has a *flux vector* $\{\Phi\}$ with components ϕ_i along the respective coordinate directions x_i. The flux is derivable from a *scalar* potential $u(\{x\})$.

$$\{\Phi\} = [C]\{u_{,i}\} \quad \text{where} \quad u_{,i} = \frac{\partial u}{\partial x_i}, \tag{22}$$

corresponds to indicial notations and [C] is the constitutive tensor. For inhomogeneous cases the flux is solved from the first statement of Eq. (8) before Eq.(22) is encountered. Hence it remains intact. This is to say that the *flux* furnishes the foundation for the renormalization procedure.

APPLICATION OF FLUX INVARIANCE

Inhomogeneous Finite Elements

The field equation of linear elasticity conforms to Eq. (22) when a linear sum of the gradient vector $\{u_{,i}\}$ is used. Especially in the finite element method:

$$\{\sigma\} = [D][B]\{U\}, \tag{23}$$

as the coefficients of the test functions are organized in the vector $\{U\}$. The strain-displacement matrix $[B]$ contains the information regarding the displacement approximation. The elasticity tensor is depicted by $[D]$. The corresponding unknown kinematic transformation $[B^*]$, for the given constitutive tensor $[D^*]$ in the case of the anisotropic body, should correspond to the same $\{\sigma\}$ field. Hence:

$$[D][B] = [D^*][B^*], \tag{24}$$

leads to an equivalent renormalized constitutive tensor $[\tilde{D}]$ such that:

$$[\tilde{D}] = [D]\left([D^*]^{-1}\right)[D], \tag{25}$$

which will enter the integral for the stiffness computation with $[B]$ for the homogeneous case along with its transpose.

Application for Biharmonic Equations

The following fourth order partial differential equation is considered:

$$\nabla^4 u(x_i) = f(x_i) \tag{26}$$

The inhomogeneous counterpart of Eq.(1) is:

$$\nabla^2\left(a(x_i)\nabla^2 u(x_i)\right) = f(x_i) \tag{27}$$

where $a(x_i)$ is the **inhomogeneous** constitutive property described as a function of the coordinates $\{x_i\}$.

According to the boundary element method, one first seeks the Green's function solution G^* for the left hand side of Eq. (2) with the Dirac's function on the right hand side:

$$\nabla^2\left(a(x_i)\nabla^2 G^*(x_i|x_i^0)\right) = \delta(x_i - x_i^0) \tag{28}$$

It will be presupposed that the corresponding Green's function, G, for the homogeneous case is available at hand. The purpose of the paper is to demonstrate how to generate G^* from G. Thus:

$$\nabla^2\left(\nabla^2 G(x_i)\right) = \delta(x_i - x_i^0) \tag{29}$$

It is sufficient for the Green's functions to satisfy the appropriate **Sommerfeld conditions of regularity at infinity** in order to be applicable in the boundary element formulation.

Hence from Eqs. (28) and (29) one obtains:

$$a(x_i)\nabla^2 G^*(x_i) = \nabla^2 G(x_i) \tag{30}$$

The Green's function, $G^{(0)}(x_i)$, for the Laplacian, ∇^2 is well known. According to the following convolution product we can then formally construct:

$$G^*(x_i) = G^{(0)}(x_i) \otimes \left(\frac{G^0(x_i)}{a(x_i)}\right) \tag{31}$$

Application for Self-adjoint Equations

For the general case of the form:

$$\mathcal{L}^{(0)}\left[a(x_i)\mathcal{L}^{(0)}[u(x_i)]\right] = f(x_i) \tag{32}$$

It will be assumed that the Green's functions, G^0 and G, for the operators $\mathcal{L}^{(0)}$ and for $\mathcal{L}$, respectively, are available. Hence Eq. (31) can still be used.

NUMERICAL EXAMPLE

The Green function $g(x)$ of a damped vibration problem has been numerically simulated as an illustration. The one-dimensional spatial variability of the inhomogeneous material property is depicted by $a(x)$.

The **computer mathematics** environment *MATHEMATICA* [7] has been used. The screen has been dumped as shown in Table 1. There are three windows therein. Window-I shows the numerical output, Window- II presents the graphs of responses. Window- III displays the actual *MATHEMATICA* code used to generate the Green function for an inhomogeneous medium $g^*(x)$.

The Green function for the homogeneous continuum **g** and the inhomogeneous distribution of material property **a**, in Window-I, have been generated as random numbers between zero and one. The operator **t** of Window-III has been used. A least square fit of all discrete data up to the fifth degree of a polynomial is carried out by the operator **fit** of Window-III. The numerical calculation involved Fourier transformation as derived in Eqs. (13) and (31). The computed values are $Re(g^*)$ and $Im(g^*)$, respectively, the real and imaginary components of the Green function that pertains to the inhomogeneous medium. These are depicted in the last two Taylor expansions in Window-I.

The numerical example is a typical case of one-parameter inhomogeneous medium where the wave speed remains unchanged as the elastic moduli and the densitv are idealized to vary proportionally.

CONCLUSIONS

The formulation for the force-response relation, according to the boundary element method, hinges on the availability of the fundamental Green's solution to the field equation in question. For media with homogeneous material properties, such functions are available in the analytical or at least in the numerical form, for a large number of problems of practical engineering interest. In the case of arbitrarily inhomogeneous constitutive property, it is only feasible to use the associated Green's functions in the numerical form. A method is provided in this paper to generate such a fundamental response from its homogeneous counterpart. Furthermore a computational strategy to employ directly an available boundary element computer program (that accepts only homogeneous constitutive fields) is sketched out herein so that numerical solutions can be generated for a general description of the inhomogeneity.

ACKNOWLEDGMENTS

This research has been sponsored by the Alexander von Humboldt Foundation, West Germany and the National Center for Earthquake Engineering Research, USA.

REFERENCES

[1] Brebbia, C. A. and Walker, S., Boundary Element Technique in Engineering, Newnes-Butterworth, London, U.K., 1980.

[2] Dasgupta, G., "Stochastic Strain - Displacement Matrices for Finite Elements with Random Constitution, " Proceedings, ICOSSAR '89, San Francisco, ASCE publication, 1989.

[3] Stakgold, I., Green's Functions and Boundary Value Problems, John Wiley, New York, 1979.

[4] Zienkiewicz, O. C., The Finite Element Method, Third Edition, McGraw-Hill, New York, NY, 1977.

[5] Dauson, D. J., "BEASY " - A Boundary Element Analysis System," Proceedings, Fifth International Conference, Boundary Elements, Hiroshima, Japan, 1983, Springer Verlag, 1983.

[6] Dasgupta, G., "Validity of Almansi Theorems for Anisotropic Boundary Elements, " Engineering Analysis, vol. 5, No. 2, June 1988, pp. 89 - 94.

[7] Wolfram, S. Mathematica A System for Doing Mathematics by Computer, Addison Wesley, NY, 1988.

Table 1: NUMERICAL EXAMPLE

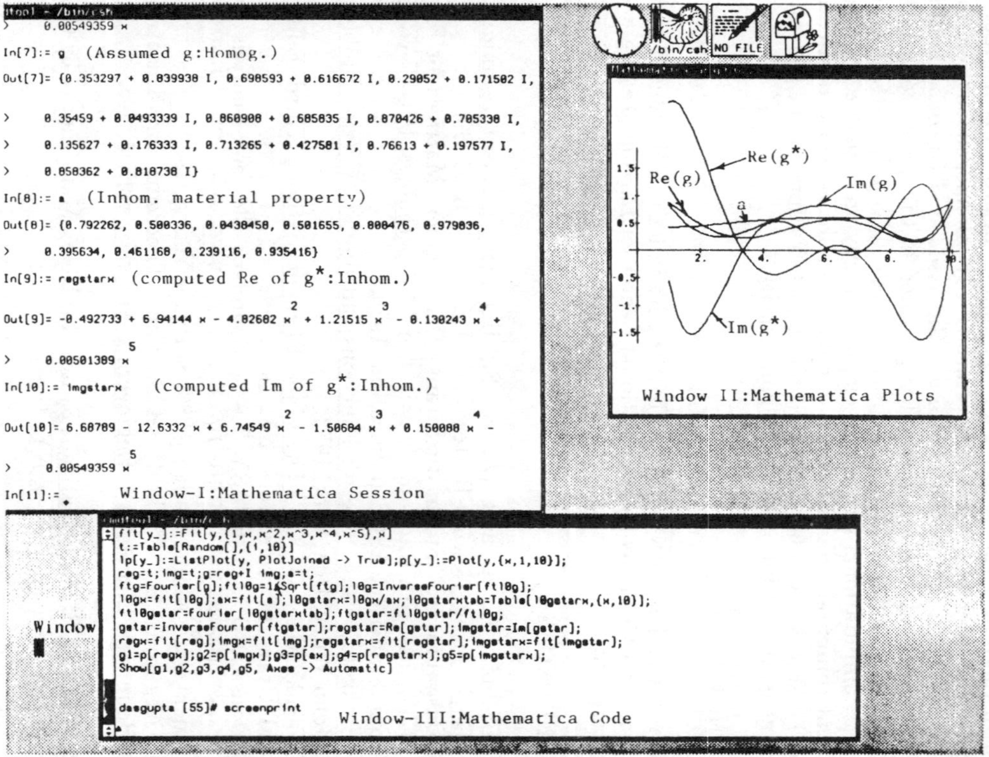

A New Hybrid Displacement Variational Formulation of BEM for Elastostatics

T.G.B. DeFigueredo* and C.A. Brebbia**

*Department of Civil Engineering, University of Brasilia, Brazil: presently at Computational Mechanics Institute, Southampton, UK

**Computational Mechanics Institute, Southampton, UK

ABSTRACT

A new variational formulation for boundary elements using a hybrid displacement functional is presented. The formulation is expressed in terms of domain and boundary variables but the functions in the domain are afterwards transformed to the boundary. The resulting equations involve only boundary variables, as in the case of conventional boundary elements and, additionally, the approach gives symmetric matrices which can be easily combined with finite elements.

Numerical results for some typical elastostatics problems have shown the accuracy of the new approach and its convergence properties.

1 Introduction

The search for new boundary element formulations is important as they have numerous applications in engineering practice, particularly for coupling to other numerical techniques or when trying to achieve better accuracy and convergence for some of the variables. This paper presents a new variational formulation of boundary elements, which results in a hybrid elasticity model that produces symmetric matrices.

Recent papers on alternative finite or boundary element formulations have concentrated on the development of hybrid stress models. Schnack[1] has used the boundary element method to generate a hybrid stress finite element model, which gives rapid convergence of the results and accurate solution for stress concentration problems. Other authors have developed variational formulations based on generalized principles of the type proposed by Hellinger-Reissner[2] and Wu-Washizu[3] in which the field and boundary variables are taken to be independent of each other. These principles have also been used in finite elements, mainly by Pian and others[4], to derive hybrid and mixed methods. Dumont[5] has presented a hybrid stress boundary element formulation based on Hellinger-Reissner's principle with stresses in the domain and displacements on the boundary as independent functions. Polizzotto[6] has proposed two other variational formulations. The first one is based on the classical Washizu and Reissner principles. The other is based on a new boundary principle expressed in terms of the unknown boundary displacements and tractions.

The method presented in this paper has been used by the authors in reference [7] for the solution of potential problems with excellent results. It is based on a modified functional using three independent variables, i.e., displacements and tractions on the boundary and displacements inside the domain. This approach uses classical fundamental solutions to interpolate the displacements in the domain and thus allowing for the transfer of the domain integrals to the boundary. The resulting system of equations is written in terms of boundary unknowns. By rearranging them one obtains a final system of equations in terms of boundary displacements only. Furthermore, this system has the advantage of being symmetrical.

The method has been validated by comparing results against analytical solutions and those obtained using the conventional boundary element method. The numerical values demonstrate that the results are accurate on the boundary as well as at internal points and they converge towards the analytical solution.

2 Basic Relations in Elastostatics

Let us consider a domain Ω bounded by a closed surface Γ which consists of a homogeneous linear elastic and isotropic material. The elastostatics problems can be represented by the following equations:

- *Compatibility equations*

  If only small deformations are supposed to occur, the strain tensor ϵ_{ij} can be given in terms of the displacement vector u_i:

$$\epsilon_{ij} = \frac{1}{2}(u_{i,j} + u_{j,i}). \tag{1}$$

- *Equilibrium conditions*

  The stress tensor σ_{ij} must satisfy the equation:

$$\sigma_{ij,i} + b_j = 0. \tag{2}$$

  The relationship between the stress components σ_{ij} and the traction vector t_i can be written as:

$$t_i = \sigma_{ji} n_j, \tag{3}$$

  where n_j are the components of the outward unit normal to the surface at the point where t_i acts.

- *Constitutive equations*

  Hooke's law is formulated in terms of shear modulus G and Poisson's ratio ν, relating the stress tensor components σ_{ij} and the strain components ϵ_{ij}:

$$\sigma_{ij} = C_{ijkl}\epsilon_{kl}, \tag{4}$$

where

$$C_{ijkl} = \frac{2G\nu}{1-2\nu}\delta_{ij}\delta_{kl} + G(\delta_{il}\delta_{jk}). \tag{5}$$

- *Boundary conditions*

The boundary conditions for displacements and tractions are as follows:

$$essential\ conditions: \quad u_i = \overline{u}_i \quad on\ \Gamma_1, \tag{6}$$

$$natural\ conditions: \quad t_i = \overline{t}_i \quad on\ \Gamma_2, \tag{7}$$

where $\Gamma = \Gamma_1 + \Gamma_2$ and the bar indicates that those values are known.

3 Modified Variational Principle

The hybrid boundary model proposed in this paper is based on a modified variational principle. The functions assumed to be independent are:

- displacement field in the domain, $\mathbf{u}$, with u_i components;
- boundary displacement field, $\tilde{\mathbf{u}}$, with $\tilde{u}_i$ components;
- boundary tractions, $\tilde{\mathbf{t}}$, with $\tilde{t}_i$ components.

The corresponding variational functional, Π_{HB}, is defined as follows:

$$\Pi_{HB} = \int_\Omega \left(\frac{1}{2}\, u_{i,j}\, C_{ijkl}\, u_{k,l} - b_i u_i\right) d\Omega - \int_\Gamma \tilde{t}_i(u_i - \tilde{u}_i)\, d\Gamma - \int_{\Gamma_2} \overline{t}_i \tilde{u}_i\, d\Gamma. \tag{8}$$

In the above equation the boundary displacement $\tilde{u}_i$ satisfies the essential boundary conditions, i.e., $\tilde{u}_i = \overline{u}_i$ on Γ_1.

By carrying out the variations it can be shown that:

$$\delta\Pi_{HB} = \int_\Omega (\sigma_{ij,j} - b_i)\delta u_i\, d\Omega - \int_\Gamma (t_i - \tilde{t}_i)\delta u_i\, d\Gamma - \int_\Gamma (u_i - \tilde{u}_i)\delta\tilde{t}_i\, d\Gamma - \int_{\Gamma_2} (\tilde{t}_i - \overline{t}_i)\delta\tilde{u}_i\, d\Gamma, \tag{9}$$

where σ_{ij} and t_i are the stress tensor and the traction vector, respectively, which are functions of the u_i displacement in the domain, through equations (1), (3) and (4).

The vanishing of $\delta\Pi_{HB}$ for arbitrary variations δu_i in Ω , $\delta\tilde{t}_i$ and $\delta\tilde{u}_i$ in Γ, with $\delta\tilde{u}_i = 0$ on Γ_1, gives the following Euler equations:

$$\sigma_{ij,j} - b_i = 0 \qquad in\ \Omega. \tag{10}$$

$$t_i - \tilde{t}_i = 0 \qquad on\ \Gamma. \tag{11}$$

$$u_i - \tilde{u}_i = 0 \qquad on\ \Gamma. \tag{12}$$

$$\tilde{t}_i - \bar{t}_i = 0 \qquad on\ \Gamma_2. \tag{13}$$

Consequently the solution of the problem is now given in terms of the functions u_i, $\tilde{u}_i$ and $\tilde{t}_i$ which make Π_{HB} stationary.

4 Discretization of the Functional

By assuming that equations (1), (3) and (4) are identically satisfied in the domain and that the displacement boundary condition is satisfied by $\tilde{u}_i$ equation (8) gives, after integrating by parts:

$$\Pi_{HB} = \int_\Gamma \frac{1}{2}\, u_i t_i \; d\Gamma - \int_\Gamma \tilde{t}_i (u_i - \tilde{u}_i)\; d\Gamma - \int_{\Gamma_2} \bar{t}_i \tilde{u}_i \; d\Gamma - \int_\Omega b_i u_i \; d\Omega - \int_\Omega \frac{1}{2}\, \sigma_{ij,j} u_i \; d\Omega. \tag{14}$$

The functions u_i, $\tilde{u}_i$ and $\tilde{t}_i$ will now be approximated as the product of known functions by unknown parameters. In matrix form they can be written as:

$$\mathbf{u} = \mathbf{U}^{*T}\mathbf{s} \qquad in\ \Omega \tag{15}$$

$$\tilde{\mathbf{u}} = \boldsymbol{\Phi}^T\mathbf{d} \qquad on\ \Gamma \tag{16}$$

$$\tilde{\mathbf{t}} = \boldsymbol{\Psi}^T\mathbf{p} \qquad on\ \Gamma \tag{17}$$

where $\mathbf{U}^*$ is a matrix whose coefficients are values of the fundamental solution at a point in the x_i coordinate directions due to a unit load acting at a point on the boundary in one of the x_i directions[8,9]; $\boldsymbol{\Phi}$ and $\boldsymbol{\Psi}$ are matrices whose terms are interpolation functions of the type used in finite or boundary elements; s, d and p are vectors of unknown parameters and T denotes transpose.

As u is expressed by equation (15), the last integral in equation (14) vanishes if one excludes from the domain Ω the singularities occurring at the source points on the boundary. These singularities will be considered when evaluating the boundary integrals.

A substitution of equations (15), (16) and (17) into equation (14) gives the functional Π_{HB} in terms of the unknown vectors s, d and p, i.e.:

$$\Pi_{HB} = \frac{1}{2}\,\mathbf{s}^T\mathbf{F}\mathbf{s} - \mathbf{p}^T\mathbf{G}^T\mathbf{s} + \mathbf{p}^T\mathbf{L}\mathbf{d} - \mathbf{d}^T\overline{\mathbf{T}} - \mathbf{s}^T\mathbf{B}, \tag{18}$$

where the matrices $\mathbf{F}$, $\mathbf{G}$, $\mathbf{L}$, $\overline{\mathbf{T}}$ and $\mathbf{B}$ are defined as:

$$\mathbf{F} = \int_\Gamma \mathbf{U}^* \mathbf{T}^{*T} \, d\Gamma,$$

$$\mathbf{G} = \int_\Gamma \mathbf{U}^* \mathbf{\Psi}^T \, d\Gamma,$$

$$\mathbf{L} = \int_\Gamma \mathbf{\Psi} \mathbf{\Phi}^T \, d\Gamma,$$

$$\overline{\mathbf{T}} = \int_{\Gamma_2} \mathbf{\Phi} \bar{\mathbf{t}} \, d\Gamma,$$

$$\mathbf{B} = \int_\Omega \mathbf{U}^* \mathbf{b} \, d\Omega.$$

$\mathbf{T}^*$ is a matrix whose terms are the tractions obtained from the displacements $\mathbf{U}^*$, corresponding to the fundamental solution[9] and the coefficienrs of $\mathbf{b}$ are body forces.

5 Final System of Equations

Since $\mathbf{u}$, $\tilde{\mathbf{u}}$ and $\bar{\mathbf{t}}$ are independent fields, the stationary value of Π_{HB} with respect to them implies:

$$\mathbf{F}\mathbf{s} - \mathbf{G}\,\mathbf{p} - \mathbf{B} = 0 \tag{19}$$

$$-\mathbf{G}^T\mathbf{s} + \mathbf{L}\mathbf{d} = 0 \tag{20}$$

$$\mathbf{L}^T\mathbf{p} - \overline{\mathbf{T}} = 0 \tag{21}$$

By substituting the values of s and $\mathbf{p}$ from equations (20) and (21), respectively, into equation (19) the final systems of equations for the solution of the problem is obtained as shown below:

$$\mathbf{K}\mathbf{d} - \mathbf{Q} = 0 \tag{22}$$

where

$$\mathbf{K} = \mathbf{R}^T\mathbf{F}\mathbf{R} \tag{23}$$

$$\mathbf{R} = \mathbf{G}^{-1^T}\mathbf{L} \tag{24}$$

$$\mathbf{Q} = \overline{\mathbf{T}} + \mathbf{R}^T\mathbf{B} \tag{25}$$

Notice that $\mathbf{K}$ is a stiffness matrix which is symmetric since the matrix $\mathbf{F}$ is symmetric.

The evaluation of the matrix $\mathbf{F}$ involves boundary integrals only and is the more critical step in this approach, as integrations of singular functions are required. Accurate and appropriate numerical integration schemes have to be used for different types of singularities that occur. However, the main diagonal of matrix $\mathbf{F}$ can be calculated by applying the rigid body motion technique and, therefore, avoiding the evaluation of hyper singular integrals.

The nodal displacements and tractions on the boundary are obtained by solving equation (22) as it is usually done in any stiffness method. However, the solution now involves only boundary unknowns and the necessary integrations to obtain the elements of $\mathbf{F}$, which produces the stiffness matrix $\mathbf{K}$, are performed merely along the boundary.

6 Displacements and Stresses at Internal Points

After calculating the boundary displacements $\mathbf{d}$ the displacements $\mathbf{u}$ and the stresses σ at internal points can be easily evaluated, without further integrations.

The internal displacement vector $\mathbf{u}$ is calculated by replacing the value of s from equation (20) into equation (15) , i.e.:

$$\mathbf{u} = \mathbf{U}^{*T}\mathbf{R}\mathbf{d}, \tag{26}$$

where $\mathbf{R}$ is defined by equation (24).

Similarly the stress tensor σ at any internal point is obtained by:

$$\sigma = \sigma^{*T}\mathbf{R}\mathbf{d}, \tag{27}$$

where the coefficients of the matrix σ^* are the stresses obtained from the displacements $\mathbf{U}^*$, corresponding to the fundamental solution[9]. Therefore, no numerical differentiation is necessary as the components of σ are known from the theory of elasticity.

7 Test Cases

Two classical elasticity problems are solved by the present approach and the results are compared to both conventional BEM and exact solutions.

7.1 Example 1

The case of a hollow cylinder under internal pressure is presented as a first numerical test (see figure (1)). This problem has been used by several authors [8,11] as a benchmark because an analytical solution is available[10]. Due to the symmetry of the problem only one quarter of the cylinder need to be modelled.

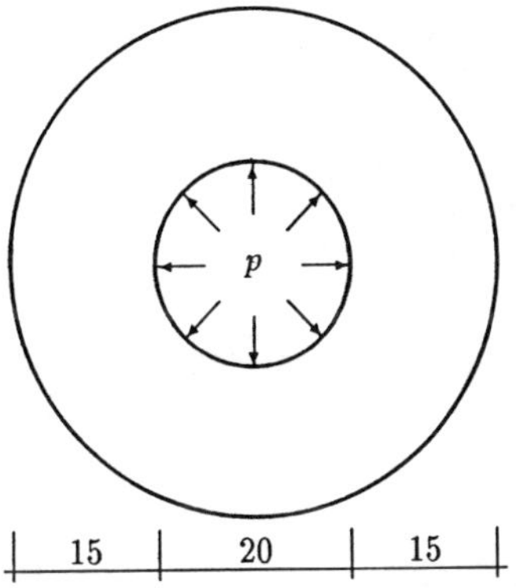

$p = 100$

$G = 80,000$

$\nu = 0.25$

Figure 1. Hollow cylinder under internal pressure: definition of the problem.

To assess the accuracy and convergence of the method three different meshes has been used, as shown in figure (2).

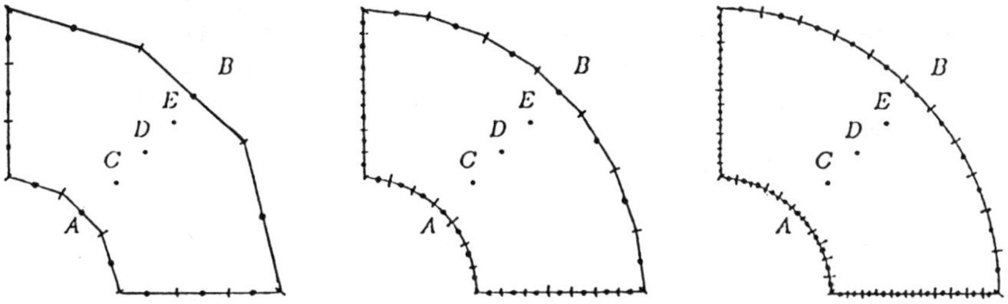

12 elements mesh | 28 elements mesh | 44 elements mesh

Figure 2. Different Boundary Element Discretizations.

Results for displacements on the boundary and at internal points are presented in table (1) and stresses at internal points are shown in table (2).

Point	Exact values	Discretization 12 elements	28 elements	44 elements
A	.80357	.78838	.80320	.80367
B	.44643	.44058	.44568	.44623
C	.62297	.61073	.62159	.62266
D	.52934	.52332	.52848	.52918
E	.47663	.50231	.47741	.47668

Table 1. Radial displacements on the boundary and at internal points (figure (2)).

Point	Exact values	Discretization 12 elements	28 elements	44 elements
C	82.0150	87.6388	82.1176	81.9496
D	57.9203	55.5943	57.9576	57.9294
E	45.4111	31.8691	43.8838	45.2755

Table 2. Circumferential stresses at internal points (see figure (2)).

Results for displacements are in all cases accurate for internal as well as for boundary points. The internal stresses, however, show considerable error for points close to the boundary when coarse meshes are used. This is to be expected as they are calculated as a superposition of fundamental solutions of higher order of singularity than the ones used for the displacements (equations (26) and (27)). The results improved considerably (table 2) when the boundary mesh is refined in the vicinity of the internal point. It is recommended that the ratio between the length of the element, closest to the internal point, and the distance from that point to the boundary be no greater than one.

7.2 Example 2

The second example considered here is the classical stress concentration problem of a square plate of width $2l$, with a small central circular hole of radius a, as depicted in figure (3). The plate is subjected to a uniform tensile stress σ. The ratio of the diameter of the hole to the width of the plate is taken to be $a/l = 0.1$ and the Poisson's ratio for the material is $\nu = 0.25$.

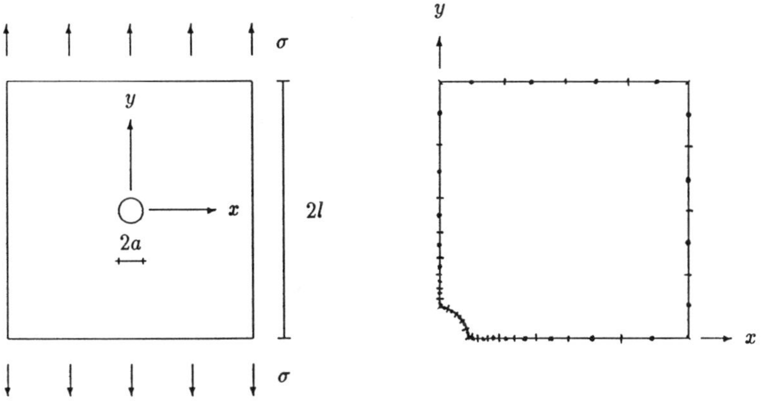

Figure 3. Square plate with hole: geometry, boundary conditions and discretization.

Figure (4) shows the variation of normal stresses in y direction along the x axis with distance from the hole. Both the conventional BEM and the hybrid BEM solutions were obtained by using the same 29 boundary constant elements mesh, shown in figure (3). K_t is the stress concentration factor defined as $K_t = \sigma_{yy}/\sigma$.

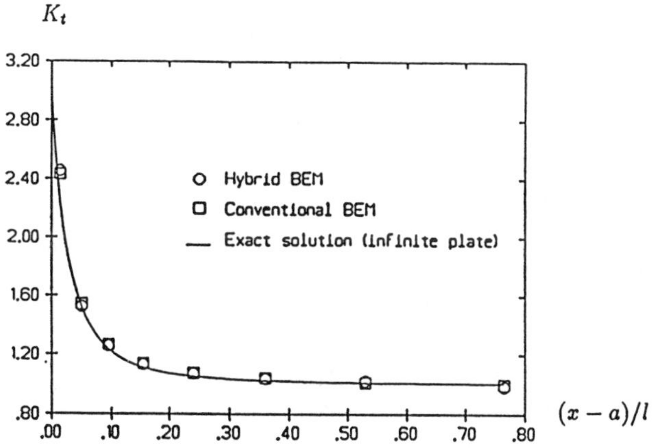

Figure 4. Variation of stresses with distance from the hole.

The analytical solution curve corresponds to the case of an infinite plate and is taken from Timoshenko[10] which gives a stress concentration factor of 3.000. This value increases slightly to 3.024 for the finite width plate (see Reference [12]).

The results for the hybrid and conventional BEM are in good agreement and they are both accurate when compared with the exact solution. Although the constant elements used here are not appropriate for modelling problems where the solution varies rapidly, the numerical results follows the analytical curve and simple extrapolation of the boundary element values give an accurate estimate of the stress concentration factor.

8 Conclusions

A new type of boundary element formulation has been presented in this paper. It is based on a hybrid elasticity model that produces symmetric matrices and involves three types of independent variables, i. e. displacements and tractions on the boundary and displacements inside the domain. By rearranging the resulting governing equations one obtains a system in terms of boundary displacements only and a symmetric stiffness matrix.

Numerical examples have shown the accuracy and convergence of the results. The solution is very accurate for tractions and displacements on the boundary. The calculation of displacements and stresses at internal points does not demand the evaluation of any integrals as in the conventional BEM, although the results are more sensitive to the proximity of the interior points to the boundary.

ACKNOWLEDGEMENTS

The first author wishes to acknowledge the financial support of the Brazilian National Council for the Scientific and Technological Development - CNPq.

REFERENCES.

1. Schnack, E., A Hybrid BEM Model, Int. Jour. Num. Meth. Engng. , vol. 24, pp. 1015-1025, 1987.
2. Reissner, E., On a Variational Theorem in Elasticity, Jour. Math. Phys., vol. 29, pp. 90-95, 1950.
3. Washizu, K., Variational Methods in Elasticity and Plasticity, 3rd Edition, Pergamon Press, Oxford, 1982.
4. Pian, T.H.H., Basis of Finite Element Methods for Solid Continua, Int. Jour. Num. Meth. Engng., vol. 1, pp. 3-28, 1969.
5. Dumont, N.A., The Hybrid Boundary Element Method, Proc. 9th Int. Conf. on BEM, Stuttgart, Computational Mechanics Publications, Southampton and Springer Verlag, Berlin, 1987.
6. Polizzotto, C., Variational Principles for Boundary Element Formulations in Structural Mechanics, Proc. 10th Int. Conf. on BEM, Southampton, Computational Mechanics Publications, Southampton and Springer Verlag, Berlin, 1987.

7. DeFigueiredo, T.G.B. and Brebbia, C.A., A Hybrid Displacement Variational Formulation of BEM, Proc. 10th Int. Conf. on BEM, Southampton, Computational Mechanics Publications, Southampton and Springer Verlag, Berlin, 1988.

8. Brebbia, C.A., The Boundary Element Method for Engineers, Pentech Press, London, 1978.

9. Brebbia, C.A., Telles, J.C.F. and Wrobel, L.C., Boundary Element Techniques Theory and Applications in Engineering, Springer-Verlag, Berlin, 1984.

10. Timoshenko, S.P. and Goodier, J.N., Theory of Elasticity, 3rd Edition, McGraw-Hill, 1982.

11. Brebbia, C.A. and Dominguez, J., Boundary Elements An Introductory Course, Computational Mechanics Publications, Southampton and McGraw-Hill, NY, 1989.

12. Roark, R.J. and Young, W.C., Formulas for Stress and Strain, 5th Edition, McGraw-Hill, London, 1975.

On Two Different Methods for Transforming Domain Integrals to the Boundary

C.A. Brebbia

Computational Mechanics Institute, Wessex Institute of Technology, Ashurst Lodge, Ashurst, Southampton SO4 2AA

1 Introduction

Since 1978 [1] the Boundary Element Method has grown considerably and been applied to solve a wide range of engineering problems, particularly those involving linear analysis. Further increase in the number of applications of the method has been hampered by the need to operate with relatively complex fundamental solutions or the difficulties encountered when trying to extend the BEM to non-linear and time dependent problems [2][3].

In many of the above cases it is necessary to define a series of internal cells where domain integrals are computed. Although the cells do not introduce any new unknowns, the resulting integrations make the method cumbersome to use.

Two new and different approaches have been proposed at the Computational Mechanics Institute, WIT. The generalization of the concept of using particular integrals produced in 1982 the technique now called Dual Reciprocity Method (DRM) [4][5][6]. This technique has been extensively studied and applied to different problems by Brebbia, Wrobel, Nardini and Niku [7][8][9][10] amongst others. a full description of the DRM with particular reference to its applications in elastodynamics can be found in reference [11].

The second approach developed at the Institute is the Multiple Reciprocity Method (MRM) originally due to Nowak [12], and it has been applied to a range of parabolic and hyperbolic problems [13][14]. The idea in MRM is to employ a sequence of higher order fundamental solutions instead of approximating the function in the domain as is done in DRM. This permits integration by parts to be applied to each term of the sequence and as a consequence the method leads to convergent boundary only solutions of the domain integrals. The technique has been applied to solve diffusion type problems [13]. Results for the Helmholtz equation can be seen in reference [14].

This paper explains both techniques - DRM and MRM - with reference to Poisson's type equation for simplicity. The case of diffusion is explained in more detail to demonstrate how the methods can be applied to time dependent problems. Once this is understood it is possible to extend their applications to a wide range of time dependent problems including hyperbolic problems. The extension to stress analysis is also straightforward (see [15]).

The paper uses the same notation for the two techniques which permits us to appreciate better the relationship between DRM and MRM, and it is interesting to point out that each of them can be seen as the 'adjoint' of the other. The advantages of each approach are discussed in detail in the paper.

2 Basic Formulation

In order to illustrate the different ways in which domain integrals in BEM can be taken to the boundary, the case of Poisson's equation will be studied. Domain integrals can arise in boundary elements due to a variety of effects such as distributed body forces, nonlinearities and time dependent problems. In what follows the right hand side of the equation will be assumed to include these effects when necessary. Hence one starts by considering the following differential equation applied in the Ω domain, i.e.:

$$\nabla^2 u = b \qquad \text{in } \Omega \tag{1}$$

where u is the potential function under study and ∇^2 the Laplacian operator. Although b is at present assumed to be a known function of space later on the case of $b(u)$ will be considered.

For simplicity let only two types of boundary conditions be taken into account, i.e.:

$$\text{i) Essential conditions of the type } u = \bar{u} \text{ on } \Gamma_1 \tag{2}$$

$$\text{ii) Natural conditions such as } q = \frac{\partial u}{\partial n} = \bar{q} \text{ on } \Gamma_2 \tag{3}$$

where Γ indicates the total boundary ($\Gamma = \Gamma_1 + \Gamma_2$). The q is the derivative of the potential with respect to the normal to the boundary n. The bar means that these values are known.

The problem can be expressed in boundary integral form by using the fundamental solution to Laplace's problem, i.e. a function defined as u^* and such that it satisfies the following equation,

$$\nabla^2 u^* + \Delta^i = 0 \tag{4}$$

with $q^* = \partial u^* / \partial n$.

Δ^i represents a Dirac delta function applied at a particular point x_i. These solutions are well known for different types of potential problems and can be found in books dealing with potential theory or boundary elements such as references [1],[2] or [3].

Applying weighted residuals with u^* as the weighting function or what is the same - using reciprocity relations between the u and u^* fields, one can transform equation (1) into a boundary integral expression [3]. For the fundamental solution applied at a given point 'i' one obtains the following formula

$$c^i\, u^i + \int_\Gamma q^*\, u\, d\Gamma + \int_\Omega b\, u^*\, d\Omega = \int_\Gamma q\, u^*\, d\Gamma \tag{5}$$

Notice that all integrals are on the boundary with the exception of the one corresponding to the b term. The constant c^i depends on the type of boundary under consideration and varies between 0 and 1, with $\frac{1}{2}$ for the case of smooth boundaries.

Expression (5) is the starting point for the application of boundary elements in potential problems. One can now subdivide the boundary Γ into elements and integrate expression (5) for the case of having n different points of application of the singular solution u^*. The procedure is well known and described in detail in the classical BEM references such as [1]. The interested reader can consult also [3] for instance, which explains the formation of the boundary element matrices in full detail. The final result is a system of equations which are normally written as below

$$\mathbf{H}\,\mathbf{U} + \mathbf{D} = \mathbf{G}\,\mathbf{Q} \tag{6}$$

where $\mathbf{H}$ and $\mathbf{G}$ are the - square - influence matrices corresponding to the boundary terms in q^* and u^*, plus the addition in the case of the $\mathbf{H}$ matrices of the diagonal terms in c^i. The vector $\mathbf{D}$ represents the results obtained by integrating the domain terms for the case of a known function b. In many cases computation of the volume integral may involve unknown functions such as u or q. Even if the b term is known throughout the domain it is usually computationally more efficient to convert it into a boundary term and then integrate. $\mathbf{U}$ and $\mathbf{Q}$ are the unknowns for potential and fluxes before any boundary conditions are applied. These conditions will have the effect of reordering the system of equations (6) in such a manner that only one unknown per node (potential or flux) will exist. This gives the final system which can then be solved, i.e.:

$$\mathbf{a}\,\mathbf{X} = \mathbf{F} - \mathbf{D} \tag{7}$$

where $\mathbf{X}$ is the vector of nodal unknowns and $\mathbf{F}$ is produced by the known values of potential and fluxes.

3 The Dual Reciprocity Method

The Dual Reciprocity Method (DRM) was invented by Nardini and Brebbia [4,5] in 1982 for the solution of elastodynamic problems. It is essentially a generalized way of constructing particular solutions and it can be used to represent body force distributions as well as for solving non-linear and time dependent problems. The method can be applied to define body force distribution over the whole domain or only part as shown by Niku and Brebbia [10].

It was realized early on that the approach could be used to solve a wide variety of parabolic as well as hyperbolic time dependent problems [9]. a complete description of how the method can be applied to vibrations has been presented by Nardini and Brebbia in reference [11]. The case of transient thermal analysis has been studied in detail by

Brebbia et al [7,15,16,17,18] including the extension of the technique to deal with non-linear diffusion problems and axisymmetric problems.

The Dual Reciprocity Method starts by representing any given function b as a combination of a series of known coordinate functions f^j where $f^j = f(\xi_j, x)$ is a function between a point ξ_j and any other point x. α^j are unknown coefficients associated with each of the f^j functions. The f^j are considered to originate at 'j' different points, many of them - but usually not all - on the boundary. These points are called 'poles' and the f^j functions are the same type for all those points. The form of b is hence as follows,

$$b = \sum_{j=1}^{M} f^j \alpha^j \tag{8}$$

The next stage is to define the particular solution, corresponding to the generic function f^j, i.e.

$$\nabla^2 \hat{u}^j = f^j \tag{9}$$

The $\hat{u}^j$ field can be found by integrating the above equation. One can then compute the value of any associated variable such as $\hat{q}^j$ by differentiating the particular solution.

Equation (5) can now be written as follows,

$$c^i u^i + \int_\Gamma q^* u \, d\Gamma - \int_\Gamma q^* u^* \, d\Gamma = -\sum_{j=1}^{M} \left\{ \alpha^j \int_\Omega (\nabla^2 \hat{u}^j) u^* \, d\Omega \right\} \tag{10}$$

Each term on the right hand side of the above equation can be integrated by parts resulting in only boundary integrals. (Notice that each 'j' term involves a particular solution localized at a 'j' pole). This gives

$$c^i u^i + \int_\Gamma q^* u \, d\Gamma - \int_\Gamma q^* u^* \, d\Gamma = \tag{11}$$
$$+ \sum_{j=1}^{M} \left\{ \alpha^j \left[c^j \hat{u}^j + \int_\Gamma q^* \hat{u}^j \, d\Gamma - \int_\Gamma u^* \hat{q}^j \, d\Gamma \right] \right\}$$

This formula has only boundary integrals but the right hand side is not found by using only one particular solution but by the addition of a series of terms, each of them representing the effect of a particular solution localized at a 'j' pole.

After applying the usual boundary element discretization equation (11) gives rise to the following system,

$$\mathbf{H\,U} - \mathbf{G\,Q} = [\mathbf{H}\,\hat{\mathbf{U}} - \mathbf{G}\hat{\mathbf{Q}}]\,\boldsymbol{\alpha} \tag{12}$$

or simply

$$\mathbf{H\,U} - \mathbf{G\,Q} = \mathbf{S}\;\boldsymbol{\alpha} \tag{13}$$

$$\text{where } \mathbf{S} = \mathbf{H}\,\hat{\mathbf{U}} + \mathbf{G}\,\hat{\mathbf{Q}} \tag{14}$$

$\hat{\mathbf{U}}$ and $\hat{\mathbf{Q}}$ are matrices N x M each for the boundary solution. (M is the number of points where the function 'j' has been applied, N is the number of boundary unknowns as usual and L the number of internal nodes. Hence $M = N + L$). The columns of these matrices represent the values of the $\hat{u}^j$ and $\hat{q}^j$ functions at the different nodes for the case of the f^j function acting at a particular 'j' pole.

Notice that the α coefficients are different from the values of the b function at the points under consideration. Both are related through equation (15). This gives a M x M matrix F which can be inverted, i.e.

$$\mathbf{B} = \mathbf{F}\,\boldsymbol{\alpha} \tag{15}$$

where $\mathbf{B}$ vector represents the magnitudes of the b functions at the poles. $\mathbf{F}$ is a matrix whose elements are the values of f^j at all points for each position 'j' of the function f. Hence one can invert (15) to find α vector, i.e.

$$\boldsymbol{\alpha} = \mathbf{F}^{-1}\,\mathbf{B} = \mathbf{E}\,\mathbf{B} \tag{16}$$

where $\mathbf{E} = \mathbf{F}^{-1}$.

Equation (20) can now be written as,

$$\mathbf{H}\,\mathbf{U} - \mathbf{G}\,\mathbf{Q} = (\mathbf{S}\,\mathbf{E})\mathbf{B} \tag{17}$$

The **H, G** and **S** matrices are functions of the geometry of the problem, the position of nodes and poles, type of interpolation and the $\hat{u}^j$, $\hat{q}^j$ fields of the particular solutions.

Choice of Functions

The most important consideration in the computational implementation of the DRM is the choice of appropriate interpolation functions f^j. After a series of numerical experiments, Brebbia and Nardini [9][11] proposed using 'conical' functions of the type distance between the points of application of the function ξ, and any given point x, such as

$$f^j = r(\xi_j, x) \tag{18}$$

This gives a very simple type of u^j function for the Laplace operator, namely,

$$\hat{u}^j(x) = \frac{1}{9}\,r^3 \tag{19}$$

In addition those authors recommended adding a suitably chosen constant for completeness, which can then be incorporated through the following function,

$$f^j = 1 \tag{20}$$

This gives the following particular solution for the Laplace's equation

$$\hat{u}^j = \frac{r^2}{4} \tag{21}$$

The introduction of one more unknown will require setting up another equation in terms of α, which can be done by defining an internal degree of freedom or 'pole'. In general the introduction of more degrees of freedom is always recommended to obtain better results when the b function is difficult to represent as a function of the boundary values only.

Other types of function proposed by Nardini and Brebbia can be seen in reference [9]. They include localized harmonic functions and polynomials in terms of the coordinates, but the best results were always reported using 'conical' functions.

For axisymmetric cases the solution depends not only on r but also on the distance from the source and field points to the axis of revolution. Because of this Wrobel, Telles and Brebbia [18] have proposed the following function,

$$f^j = r\left(1 - \frac{R_j}{4R}\right) \tag{22}$$

where R_j is the distance from the different poles to the axis of revolution and R is the distance from any point on Γ (or Ω) to the same axis. The $\hat{u}$ and $\hat{q}$ functions are the same type as for three dimensional analysis.

Example 1

Solid Cylinder with Constant Heat Transfer Coefficient

The DRM was introduced into a commercial package, i.e. the BEASY analysis code [19] to represent heat diffusion problems including the case of material properties which vary with temperature. The present example considers the material properties to be constant and analyses a solid cylinder subject to a convective boundary condition with a heat transfer coefficient constant on time.

The form of equation (1) for transient heat transfer problems is

$$\nabla^2 u = \frac{1}{\kappa}\,\dot{u} \tag{a}$$

where u is now the temperature, κ the diffusivity and $\dot{u}$ represents the derivative of temperature with respect to time. Convective boundary conditions are of a mixed type and give the following expression for the flux,

$$q = h(u - u_f) \qquad \text{on } \Gamma_c \tag{b}$$

where h is the heat transfer coefficient, u_f is the ambient temperature. Γc represents the part of the boundary where convective boundary conditions are applied.

Notice that the domain term b in (1) is now replaced by $\frac{1}{\kappa}\,\dot{u}$ and hence the corresponding integral can be written as,

$$\int_\Omega u^*(\frac{1}{\kappa}\,\dot{u})\,d\Omega \tag{c}$$

As κ is assumed to be a constant here one can take it out of the integral and expand $\dot{u}$ in a similar way to the procedure used for b, i.e.

$$\dot{u} = \sum_{j=1}^{M} f^j(x)\,\dot{\alpha}^j(t) \tag{d}$$

Notice that the f^j function depends on space and the $\dot{\alpha}$ coefficients on time.

Proceeding as before one finds that the governing integral equation for DRM can now be written as,

$$c^i\,u^i + \int_\Gamma q^*\,u\,d\Gamma - \int_\Gamma u^*\,q\,d\Gamma = \tag{e}$$
$$= \sum_{j=1}^{M} \frac{1}{\kappa}\,\{c^j\,\hat{u}^j + \int_\Gamma q^*\,\hat{u}^j\,d\Gamma - \int_\Gamma u^*\,\hat{q}^j\,d\Gamma\}\,\dot{\alpha}^j$$

After applying the normal BE approximations the above statement produces the following system of equations,

$$\mathbf{H}\,\mathbf{U} - \mathbf{G}\,\mathbf{Q} = \frac{1}{\kappa}(\mathbf{H}\,\hat{\mathbf{U}} - \mathbf{G}\,\hat{\mathbf{Q}})\,\boldsymbol{\alpha} \tag{f}$$

After evaluating expression (d) and all poles one can find the relationship

$$\dot{\mathbf{U}} = \mathbf{F}\,\dot{\boldsymbol{\alpha}} \qquad (g)$$

and its inverse gives

$$\dot{\boldsymbol{\alpha}} = (\mathbf{F}^{-1})\dot{\mathbf{U}} = \mathbf{E}\,\dot{\mathbf{U}} \qquad (h)$$

which finally results in a system similar to those obtained in domain methods such as finite differences or finite elements, i.e.,

$$\mathbf{C}\,\dot{\mathbf{U}} + \mathbf{H}\,\mathbf{U} = \mathbf{G}\,\mathbf{Q} \qquad (i)$$

where $\mathbf{C} = -\frac{1}{\kappa}(\mathbf{H}\,\hat{\mathbf{U}} - \mathbf{G}\,\hat{\mathbf{Q}})\mathbf{E}$.

Any standard direct integration can now be used to find the solution of equation (*i*). The present version of the BEASY system uses a simple two level time integration scheme as described in [12].

Consider the use of a solid cylinder of radius 0.10, and diffusivity 2 x 10^{-5}. The convective conditions are assumed to be constant with time. The heat transfer coefficient h is equal to 420 for an ambient temperature of 50.

The cylinder was considered as a two and three dimensional system always using the facilities of BEASY. The theoretical results were taken from [20]. The initial state at time 0 is a constant temperature of 400° everywhere in the cylinder.

The results at time 1200 are as follows:

Table 1

Case	Radial distance	Potential	Surface Flow
Two Dimensional BEM	0.	75.2	-
(BEASY)	0.10	69.1	8033
Three Dimensional BEM	0.	98.	-
(BEASY)	0.10	88.	15,700
Analytical Solution	0.	81.5	-
[18]	0.10	73.56	9912

The analytical solution was approximate and can only be taken as an indication of the order of the results. In the three dimensional case only four quadratic elements were employed on the face with convective boundary conditions, which was on one quarter of the total cylinder due to symmetry. For two dimensions four quadratic elements were used to describe the same quarter. Hence one would expect more accurate results for the two dimensional case. Notice also that the values of temperature at time 1200 are low when compared with the original temperature of 400 degrees and that the time step taken - i.e. 150 - was much larger than using FD methods.

4 Multiple Reciprocity Method

The Multiple Reciprocity Method (MRM) is a new technique of transforming domain integrals to the boundary. It has some features which are similar to the DRM but instead of approximating the source term by a set of coordinate functions a sequence of functions related to the fundamental solution is introduced in MRM. These constitute a set of higher order fundamental solutions which permits the application of Green's identity to each term of a sequence in succession. As a result, the method can lead in the limit to the exact boundary only formulation of the problem.

The MRM was introduced by Nowak [12] and extended to a series of applications by Nowak and Brebbia [13][14], including transient and Helmholtz type problems. The name MRM was proposed by Nowak and Brebbia in reference [13].

Consider again the case of Poisson's equation (1) for simplicity but with the subscript 'o' on the right hand side of the term to differentiate from other similar functions which will be generated during the solution, i.e.

$$\nabla^2 u = b_o \qquad \text{in } \Omega \tag{23}$$

where u and b_o are the potential and the source functions respectively.

The fundamental solution to Laplace's equation defined by (4) will also be called with a subscript 'o' for the reason explained above, i.e.

$$\nabla^2 u_o^* + \Delta^i = 0 \tag{24}$$

Applying reciprocity to the above relations one obtains the same expression as before i.e. equation (5), which is written again below for completeness

$$c^i u^i + \int_\Gamma q_o^* \, u \, d\Gamma + \int_\Omega b_o \, u_o^* \, d\Omega = \int_\Gamma q \; u_o^* \, d\Gamma \tag{25}$$

where $q_o^* = \partial u_o^* / \partial n$.

The domain integral in (25) can now be transformed into a series of equivalent boundary integrals. In order to do this one introduces a new function u_1^* related to the fundamental solution u_o^* by the formula

$$\nabla^2 u_1^* = u_o^* \tag{26}$$

Thus the domain integral in (25) can be expressed as follows,

$$\begin{aligned} \int_\Omega b_o \, u_o^* \, d\Omega &= \int_\Omega b_o \, (\nabla^2 u_1^*) d\Omega = \\ &= \int_\Gamma \left\{ b_o \frac{\partial u_1^*}{\partial n} - u_1^* \frac{\partial b_o}{\partial n} \right\} d\Gamma + \int_\Omega u_1^* \, \nabla^2 \, b_o \, d\Omega \end{aligned} \tag{27}$$

As the source function b_o is at present assumed to be a known function of space, one can obtain the function $\nabla^2 b_o$ analytically and hence one can define a new function b_1 such that

$$b_1 = \nabla^2 b_o \tag{28}$$

The domain integral on the right hand side of (27) can then be written in a form which is similar to the one for the previous domain integral and can be expanded in the same way, i.e.

$$\int_\Omega b_1 \, u_1^* \, d\Omega = \int_\Gamma \left\{ b_1 \frac{\partial u_2^*}{\partial n} - u_2^* \frac{\partial b_1}{\partial n} \right\} d\Gamma + \int_\Omega u_2^* (\nabla^2 \, b_1) d\Omega$$

Notice that one can now compute a b_2 function such that

$$b_2 = \nabla^2 b_1 \tag{29}$$

and continue carrying out this procedure as many times as desired.

The procedure can be generalized by introducing two sequences of functions defined by the following recurrence formula,

$$\begin{aligned} b_{j+1} &= \nabla^2 b_j \\ & \qquad\qquad \text{for } j = 0, 1, 2... \\ \nabla^2 u^*_{j+1} &= u^*_j \end{aligned} \tag{30}$$

The domain integral can then be expressed as

$$\int_\Omega b_o\, u^*_o\, d\Omega = \sum_{j=1}^{\infty} \int_\Gamma \left[b_j \frac{\partial u^*_{j+1}}{\partial n} - u^*_{j+1} \frac{\partial b_j}{\partial n} \right] d\Gamma \tag{31}$$

Introducing (32) into the original boundary integral expression one obtains the exact boundary only formulation of the problems, i.e.

$$\begin{aligned} c^i u^i + \int_\Gamma u\, q^*_o\, d\Gamma &= \int_\Gamma q\, u^*_o\, d\Gamma \\ - \sum_{j=0}^{\infty} \int_\Gamma & \left[b_j \frac{\partial u^*_{j+1}}{\partial n} - u^*_{j+1} \frac{\partial b_j}{\partial n} \right] d\Gamma \end{aligned} \tag{32}$$

The integrals can be evaluated numerically by subdividing the boundary Γ into elements as usual. As the functions b_j are known functions of space the integrals in the summation can be calculated directly. The same type of interpolations used for u and q can be applied for b_j. Thus equation (32) can be expressed in terms of the usual boundary element influence matrices now called $\mathbf{H}_o$ and $\mathbf{G}_o$ plus those matrices resulting from the use of higher order fundamental solutions, which are defined as $\mathbf{H}_{j+1}$ and $\mathbf{G}_{j+1}$ $(j = 0, 1, 2...)$ i.e.

$$\mathbf{H}_o\, \mathbf{U} - \mathbf{G}_o\, \mathbf{Q} = \sum_{j=0}^{\infty} (\mathbf{H}_{j+1}\, \mathbf{P}_j - \mathbf{G}_{j+1}\, \mathbf{R}_j) \tag{33}$$

The vectors $\mathbf{P}_j$ and $\mathbf{R}_j$ contain the function b_j and its normal derivatives respectively at the boundary nodes.

Notice that the terms of the series in the right hand side of equation (33) vanish rapidly provided that the problem has been properly scaled, (i.e. all dimensions are divided by the maximum dimension of the problem). It has been shown that the convergence of the series is very rapid in a variety of practical cases. Furthermore this convergence can be easily calculated since all the terms are known and the contribution of each of them can be evaluated. It should also be pointed out that the functions u_j^* have no singularities for $j = 1, 2...$ and thus their integration does not require any special technique.

Equation (33) is solved using standard boundary element subroutines after taking into consideration the boundary conditions.

Higher Order Fundamental Solutions

The higher order fundamental solutions required here are defined by the recurrence formula (31). This equation can be easily solved analytically when the Laplace's operator is written in terms of a cylindrical (for 2-D problems) or spherical (for 3-D cases) coordinate system. For example, for 2-D problems the general form of the function u_j^* is given by the following expression,

$$u_j^* = \frac{1}{2\pi} r^{2j} (A_j \ln r - B_j) \tag{34}$$

where r represents the distance between points. The coefficients A_j and B_j are obtained for the following recurrence relationships

$$\begin{aligned} A_{j+1} &= \frac{A_j}{4(j+1)^2} \\ B_{j+1} &= \frac{1}{4(j+1)^2} \left\{ \frac{A_j}{j+1} + B_j \right\} \end{aligned} \tag{35}$$

For $j = 0$ one obtains the classical fundamental solution, i.e. $A_o = 1$ and $B_o = 0$. Notice that formulae (35) introduce factorials into the denominators of coefficients A_j and B_j and hence guarantee their rapid convergence.

Example 3

In order to demonstrate the accuracy of the proposed technique, a simple numerical example is illustrated here. Heat is conducted through the 1-D slab being insulated at $x = 0$ and subjected to the boundary condition of the third kind (i.e. equation (b) example 1) at $x = 1$. The source term is uniform within the slab. For steady-state problems, the Multiple Reciprocity Method gives the analytical solution. For transient problems an additional domain integral has to be considered. This integral has the following form

$$\int_\Omega \dot{u}\, u_o^*\, d\Omega \tag{a}$$

Notice that as $u_o^* = \nabla^2 u_1^*$ this expression can be converted into

$$\int_\Omega \dot{u}\, (\nabla^2 u_1^*) d\Omega = -\int_\Gamma \dot{u}\, q_1^*\, d\Gamma + \int_\Gamma \dot{q}\, u_1^*\, d\Gamma + \int_\Omega u_1^*\, (\nabla^2 \dot{u}) d\Omega \tag{b}$$

Noticing that $\nabla^2 u = \dot{u}$ and $\frac{\partial}{\partial t}(\nabla^2 u) = \frac{\partial}{\partial t}(\dot{u})$, the last integral can be written as

$$\int_\Omega u_1^*(\nabla^2 \dot{u})d\Omega = \frac{\partial}{\partial t}\int_\Omega u_1^*(\nabla^2 u)d\Omega = \frac{\partial}{\partial t}\int_\Omega \dot{u}\, u_1^*\, d\Omega \qquad \text{(c)}$$

One can continue to expand these integrals using the higher order versions of the above formulae and find the following relationship

$$c^i u^i + \sum_{j=0}^{\infty}\int_\Gamma q_j^* \frac{\partial^j u}{\partial t^j}\, d\Gamma = \sum_{j=0}^{\infty}\int_\Gamma u_j^* \frac{\partial^j q}{\partial t^j}\, d\Gamma \qquad \text{(d)}$$

Discretizing these expressions using the boundary element methodology one obtains

$$\mathbf{H}_0\, \mathbf{U} + \mathbf{H}_1\, \dot{\mathbf{U}} + \mathbf{H}_2\, \ddot{\mathbf{U}} + ... = \mathbf{G}_o\, \mathbf{Q} + \mathbf{G}_1\, \dot{\mathbf{Q}} + \mathbf{G}_2\, \ddot{\mathbf{Q}} + ... \qquad \text{(e)}$$

This equation can be integrated using some approximation for the time derivatives.

Details of the formulation are given in reference [12], and some representative results for the slab are shown in Figure 1. The solid lines correspond to the analytical solution whereas the symbols stand for the numerical solution obtained by the Multiple Reciprocity Method. It is interesting to point out the high accuracy of the proposed technique.

5 Conclusions

The paper discusses two different methods to transform domain integrals to the boundary in BEM applications. The problem is defined with reference to Poisson's type equation for simplicity but can easily be extended to stress analysis and other applications. The two techniques are

i) **The Dual Reciprocity Method [DRM]** originally proposed by Nardini and Brebbia in 1982 and which approximates the domain terms in function of a series of localized particular solutions. This approximation permits the domain integrals to be converted into boundary integrals by simple integration by parts.

ii) **The Multiple Reciprocity Method [MRM]** more recently developed by Nowak and Brebbia. This technique uses high order fundamental solutions to transform domain into boundary integrals.

It is interesting to point out that the techniques can be seen as 'adjoint' of each other. The DRM has proved to give accurate results in many practical applications and has been used by many authors. The MRM is however more recent and it is only now being applied to solve different types of problem. An important advantage of the MRM over DRM is that it can guarantee convergence of the solution under certain conditions by

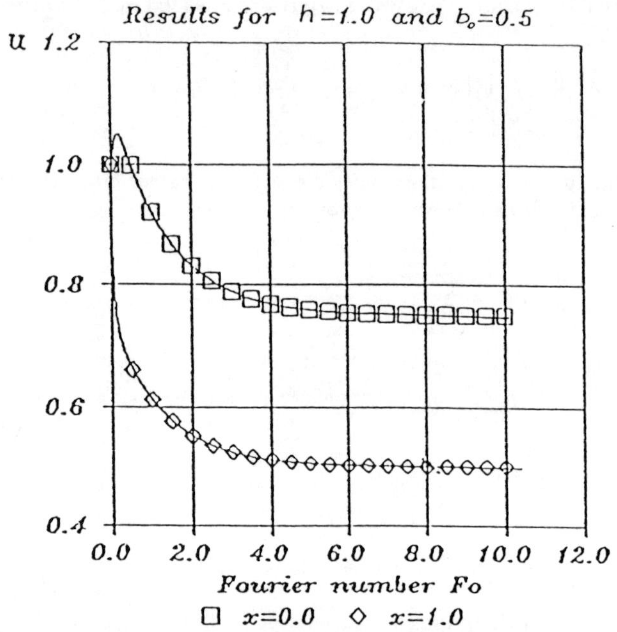

a) Potential u on the boundary as a function of Fourier number

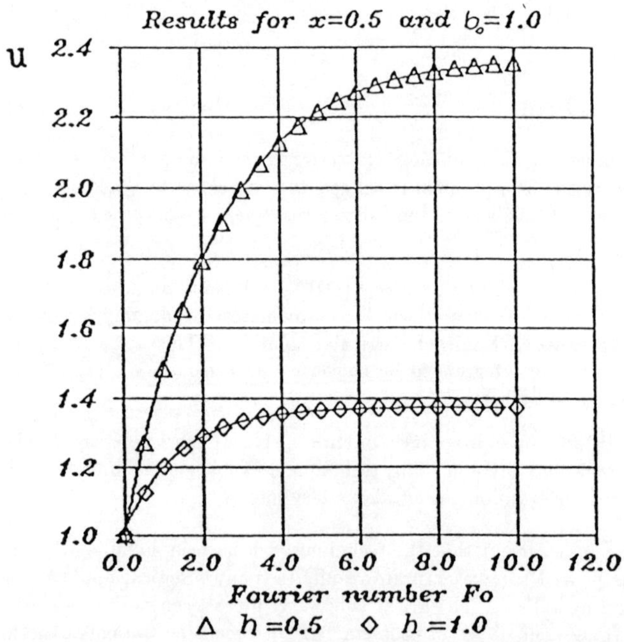

b) Potential u at the middle of the slab as a function of Fourier number

Figure 1 : Transient Heat Conduction in Infinite Slab with Constant Heat Source

introducing successive terms in the series. The accuracy of the MRM results can in this way also be controlled.

References

1. Brebbia, C.A. The Boundary Element Method for Engineers, Pentech Press, London, 1978.

2. Brebbia, C.A., Telles, J.C.F. and Wrobel, L.C. Boundary Element Techniques - Theory and Applications in Engineering, Springer-Verlag Berlin and New York, 1984.

3. Brebbia, C.A. and Dominguez, J. Boundary Elements - An Introductory Course. Computational Mechanics Publications, Southampton, and McGraw Hill, New York, 1989.

4. Nardini, D. and Brebbia, C.A. A New Approach to Free Vibration Analysis using Boundary Elements, in Boundary Element Methods in Engineering (ed. C.A. Brebbia), Springer-Verlag, Berlin and New York, 1982.

5. Brebbia, C.A. and Nardini, D. Dynamic Analysis in Solid Mechanics by an Alternative Boundary Element Approach, Int. J. Soil Dynamics and Earthquake Engineering, Vol.2, No. 4, pp. 228-233, 1983.

6. Brebbia, C.A. Solution of Time Dependent Problems using Boundary Elements. In Mathematics of Finite Elements and Applications, Academic Press, London, 1985.

7. Wrobel, L.C., Brebbia, C.A. and Nardini, D. The Dual Reciprocity Boundary Element Formulation for Transient Heat Conduction. In Proc. of the Vth Int. Conf. on FEM in Water Resources. Computational Mechanics Publications, Southampton and Boston, Springer-Verlag, Berlin, 1986.

8. Wrobel, L.C., and Brebbia, C.A. Boundary Elements for Non-Linear Heat Conduction Problems. Communications in Applied Numerical Methods, Vol. 4, pp. 617-622, 1988.

9. Nardini, D. and Brebbia, C.A. The Solution of Parabolic and Hyperbolic Problems using an Alternative Boundary Element Formulation. In Proc. of VIIth Int. Conf. on BEM in Eng. Computational Mechanics Publications, Southampton and Boston, Springer-Verlag, Berlin, 1985.

10. Niku, S.M. and Brebbia, C.A. Dual Reciprocity Boundary Element Formulation for Potential Problems with Arbitrarily Distributed Forces. Technical Note, Eng. Analysis, Vol.5, No.1, 1988.

11. Nardini, D. and Brebbia, C.A. Boundary Integral Formulations of Mass Matrices for Dynamics Analysis. In Topics in Boundary Element Research, Vol.2 (ed. C.A. Brebbia) Springer-Verlag, Berlin, and New York, 1985.

12. Nowak, A.J. Temperature Fields in Domains with Heat Sources using Boundary-only Formulations. Proc. 10th BEM Conference Southampton and Boston, 1988 (ed. C A. Brebbia) Springer-Verlag, Berlin, Vol.2, pp.233- 247, 1988.

13. Nowak, A.J. and Brebbia, C.A. The Multiple Reciprocity Method - A New Approach for Transforming BEM Domain Integrals to the Boundary. Eng. Analysis, Vol.6 No.3, 1989.

14. Nowak, A.J. and Brebbia, C.A. Solving Helmholtz Equation by Boundary Elements using the Multiple Reciprocity Method. In Computers and Experiments in Fluid Flow, (Eds. G. Carlomagno and C.A. Brebbia) Computational Mechanics Publications, Southampton and Boston Springer-Verlag, Berlin, 1989.

15. Wrobel, L.C., and Brebbia, C.A. The Dual Reciprocity Boundary Element Formulation for Non-Linear Diffusion Problems. Computer Methods in Applied Mechanics and Engg. Vol. 65, pp. 147-164, 1987.

16. Wrobel, L.C., Brebbia, C.A. and Nardini, D. Analysis of Transient Thermal Problems in the BEASY system. In BETECH/86 (Eds. J.J. Connor and C.A. Brebbia), Computational Mechanics Publications, Southampton and Boston, 1986.

17. Brebbia, C.A. and Wrobel, L.C. Non-Linear Transient Thermal Analysis using the Dual Reciprocity Method. In Boundary Element Techniques: Applications in Stress Analysis and Heat Transfer, (Eds. C.A. Brebbia and W. Venturini), Computational Mechanics Publications, Southampton and Boston, 1987.

18. Wrobel, L.C., Telles, J.C.F. and Brebbia, C.A. A Dual Reciprocity Boundary Element Formulation for Axisymmetric Diffusion Problems. In Proc. of VIIth Int. Conf. on B.E.M. in Engg., Tokyo, 1986. Computational Mechanics Publications, Southampton and Boston, Springer-Verlag, Berlin, 1986.

19. BEASY - Boundary Element Analysis System as a CIM Tool. Computer Aided Engineering Systems Handbook, Vol. 1. Computational Mechanics Publications, Southampton and Boston Springer-Verlag, Berlin, 1987.

SECTION 2 - ADAPTIVE TECHNIQUES, ERROR AND CONVERGENCE

Absolute p-Refinement of Boundary Elements for Two-Dimensional Elasticity Problems

Timothy J. Urekew and Joseph J. Rencis
Mechanical Engineering Department, Worcester Polytechnic Institute, Worcester, Massachusetts 01609, USA

ABSTRACT

A simple polynomial refinement formulation is presented for two-dimensional elasticity problems using the direct, collocation, boundary element method. A combination of analytic and numeric integrations are used in evaluating the boundary integrals for discontinuous, constant, and continuous, linear through cubic, solution distributions. The L_2 boundary error norm, relative percentage error, and solution norm ratio are used to study convergence of displacements and tractions for an L-shaped domain demonstration problem. The reference for convergence is the solution obtained from a fine, boundary element mesh.

1. INTRODUCTION

Polynomial refinement (p-refinement), in the finite (FEM) and boundary (BEM) element methods, is a means of improving an initial solution by increasing the order of the solution interpolation while holding the element size fixed. Polynomial refinement can be further broken down into absolute and relative refinement schemes. Absolute p-refinement involves increasing the solution interpolation by adding full or absolute degrees of freedom, leading to higher order solution formulations. The relative p-refinement schemes increase the order of solution interpolation by adding bubble functions or relative degrees of freedom. In this case the initial solution basis functions are either a subset of or completely different from the refinement basis functions. The families of relative refinement basis functions preserve inter-element solution continuity while modifying the solution distribution between the end nodes. Relative p-refinement is often referred to as hierarchial refinement [1, 10] in the boundary element and finite element context. Both forms of p-refinement have been thoroughly researched for the FEM [10, 13, 19, 24], and implemented into computer codes [9, 20, 21]. Only recently has polynomial convergence been researched for the BEM.

In 1985, 1986 Alarcón et al. [1, 2] studied p-convergence of two-dimensional (2-D) potential and elasticity problems using the collocation BEM. This work was carried out in an adaptive environment, using linear solution basis functions and the Peano hierarchial functions for relative refinement. Reverter et al. [17] examined refinement of potential problems using the Peano and Legendre hierarchial functions, coupled with linear solution basis functions. In 1987 Cerrolaza et al. [7] and Lera et al. [11] formulated relative p-refinement schemes for three-dimensional (3-D) potential and 2-D and 3-D potential and linear elasticity problems, respectively. Both works employed the collocation BEM, linear solution basis functions and Legendre interpolation functions for refinement. Parreira [14] studied self-adaptive p-convergence of the 2-D, Galerkin, BEM formulation for linear elasticity. This work includes convergence results for two problems, using Legendre, refinement polynomials, up to parabolic degree. S-I. Umetani [22] furthered the work of Parreira [14] by investigating relative refinement of the 2-D, Galerkin, BEM formulation for elasticity using linear through quintic Legendre polynomials. All of the publications for BEM p-refinement cited here have been developed for an adaptive environment.

The work presented in this paper uses the direct collocation BEM formulation for 2-D elasticity to examine absolute p-refinement. In Section 2 the direct collocation formulation for 2-D, linear elasticity is presented. Section 3 describes absolute refinement strategy for constant through cubic solution distributions. Section 4 establishes the convergence criteria through which the absolute refinement is studied. Lastly, in Section 5, a demonstration problem is presented to compare the proposed p-refinement strategy with the traditional refinement approach of reducing element size while holding the order of solution interpolation fixed (i.e., h-refinement).

2. BOUNDARY ELEMENT FORMULATION

The governing equations for a two-dimensional, homogeneous body assuming isotropic, linear elasticity are Navier's equations, given here as

$$\frac{G}{(1-2\nu)}\frac{\partial^2 u_j}{\partial x_i\,\partial x_j}+G\frac{\partial^2 u_i}{\partial x_j\,\partial x_j}+b_i=0\,, \qquad i=1,2\,;\; j=1,2 \qquad (1)$$

where u_j, u_i are displacement vectors and b_i is a vector of body force per unit volume. In addition, G is the shear modulus and ν is Poisson's ratio. The solution for Equation (1), in terms of the boundary integral equation (BIE), can be obtained from the direct formulation discussed in [3]. The resulting BIE can be written as

$$c_{ij}\, u_j(\xi) + \int_{\Gamma} t_{ij}^{*}(\xi,x)\, u_j(x)\, d\Gamma(x) = \int_{\Gamma} u_{ij}^{*}(\xi,x)\, t_j(x)\, d\Gamma(x) + \int_{\Omega} u_{ij}^{*}(\xi,x)\, b_j(x)\, d\Omega(x), \qquad i = 1,2\,;\; j = 1,2 \tag{2}$$

where c_{ij} is a coefficient matrix dependent upon local surface geometry, t_{ij}^{*}, u_{ij}^{*} are the fundamental solutions for Equation (1) [6], $u_j(x)$, $t_j(x)$ are displacement and traction distributions, and $b_j(x)$ is the body force per unit volume. The fundamental solutions $t_{ij}^{*}(\xi,x)$, $u_{ij}^{*}(\xi,x)$ are singular solutions to Navier's Equation (1) and are to be interpreted as tractions and displacements, respectively, in the j direction at the field point x corresponding to a unit point force acting in the i direction, applied at the source point ξ. By collocating at each of the solution nodes (source points) defined in the BE mesh, the displacement and traction coefficients in Equation (2) can be evaluated. These integrals are evaluated numerically when integrating over an element not containing the collocation point (source point), and analytically (linear geometry) when integrating over an element containing the collocation point. The coefficient matrix, c_{ij}, is $c_{ii} = 1/2$ for straight, smooth surfaces, and has been evaluated in closed form in [18] for straight, non-smooth surfaces.

A well-posed, 2-D problem will have a combination of displacements and tractions prescribed over the boundary, leaving two unknown variables per solution node. After evaluating the integrals and applying boundary conditions, Equation (2) can be rearranged into the linear system of algebraic equations,

$$A_{ij}\, x_j = b_j\,. \tag{3}$$

The unknown nodal displacements and tractions, x_j, are found by solving Equation (3).

3. ABSOLUTE REFINEMENT STRATEGY

The absolute refinement strategy requires adding new degrees of freedom to affect the distributions of $u_j(x)$ and $t_j(x)$ in Equation (2). This leads to complete higher-order, BEM solution distributions of the form

$$\hat{u}_i = \sum_{k=1}^{n} N_k\, u_k\,, \qquad i = 1,2, \tag{4}$$

$$\hat{t}_i = \sum_{k=1}^{n} N_k\, t_k\,, \qquad i = 1,2 \tag{5}$$

where n is the number of collocation points used to define the solution, and u_k, t_k are the displacements and tractions, respectively, at the collocating points. Equations (4) and (5) are the solution functions used in Equation (2) to evaluate the unsymmetric coefficient matrix, A_{ij}, in Equation (3). During mesh refinement, successively higher-order interpolation functions, N_k, are introduced until the numerical solution is within some acceptable range of error. The solution interpolation functions utilized in this work define discontinuous constant and continuous, linear through cubic distributions. These functions are in terms of a parametric coordinate system ranging from -1 to +1, and can be found in [4]. The continuous solution formulation was chosen in an effort to minimize the number of degrees of freedom to be solved for in Equation (3). Modeling difficulties associated with discontinuous boundary conditions and geometric corners are handled using the double-node concept presented in [12]. In addition to minimizing the number of degrees of freedom, the continuous formulation also enforces inter-element displacement continuity.

In this work geometry and boundary conditions are always prescribed using linear, continuous interpolation. This means that only straight elements are used for describing geometry. In addition, displacement and traction boundary conditions are prescribed as linear distributions. Restricting the order of geometry and boundary conditions was done to simplify the evaluation of the kernel functions in Equation (2) and to facilitate the extrapolation of boundary data corresponding to the new degrees of freedom that are generated during p-refinement.

The refinement strategy used in this work is uniform as opposed to the localized refinement capability of adaptive methods. It is interesting to note that the order of the final system of algebraic equations in Equation (3) is the same as that obtained from a uniform, relative refinement approach. This is because of the continuous element formulation. The relative refinement approach, however, is more computationally efficient in that all of the A_{ij} matrix coefficients from Equation (3), for lower order analyses, are used in each, successive, higher-order refinement. In the case of absolute refinement, the initial A_{ij} matrix in Equation (3) must be partially recalculated for adaptive schemes, or completely recalculated for uniform refinement. The proposed absolute refinement strategy does, however lend itself to adaptive implementation and also eliminates the difficult question of where to collocate when adding relative degrees of freedom [1, 2, 7, 8, 11, 22]. All of the convergence studies in this work utilize uniform p-refinement.

4. CONVERGENCE GAGING PARAMETERS

Convergence of the p-refinement strategy presented in Section 3, is studied on the local and global levels through the use of error indicators and estimators. Error indicators provide local or element level information about the quality of the numerical solution, relative to the exact solution. Error estimators provide global information on the quality of the numerical solution, relative to the exact solution. Typically the exact solution is not available for general, 2-D elasticity problems, therefore, an approximation to the exact must be used. In this work the exact solution is taken as the numerical solution obtained from a fine, uniform mesh of quadratic elements using BEASY [5].

The L_2 boundary error norm is used to quantify errors in traction and displacement, on both the local (element) and global levels. The L_2 boundary norm has been selected for reasons stated in [16], and is defined here, for each element, j, as

$$\| e_{u_i} \|_j = \left(\int_{\Gamma_j} (e_{u_i})^T (e_{u_i})\, d\Gamma \right)^{1/2}, \qquad i = 1,2 \tag{6}$$

for displacement components i and

$$\| e_{t_i} \|_j = \left(\int_{\Gamma_j} (e_{t_i})^T (e_{u_i})\, d\Gamma \right)^{1/2}, \qquad i = 1,2 \tag{7}$$

for traction components i. The error functions e_{u_i} , e_{t_i} associated with displacement components i and traction components i, respectively, are defined as

$$e_{u_i} = u_i - \hat{u}_i, \qquad i = 1,2, \tag{8}$$

$$e_{t_i} = t_i - \hat{t}_i, \qquad i = 1,2 \tag{9}$$

where u_i, t_i represent the exact solution of displacement and traction, respectively, and $\hat{u}_i$, $\hat{t}_i$ represent the p-refined, BEM solution of displacement and traction, respectively. Equations (5) through (9) are general expressions that can be used for studying errors in each of the orthogonal, i, components of the displacement and traction vectors. The error norms associated with the resultant displacement and traction vectors for each element, j, are defined as

$$\| e_u \|_j = \left(\| e_{u_1} \|_j^2 + \| e_{u_2} \|_j^2 \right)^{1/2}, \tag{10}$$

$$\| e_t \|_j = \left(\| e_{t_1} \|_j^2 + \| e_{t_2} \|_j^2 \right)^{1/2}. \tag{11}$$

Exact boundary solution norms are also calculated for each element, j, from

$$|| u_i ||_j = \left(\int_{\Gamma_j} u_i^2 \, d\Gamma \right)^{1/2}, \qquad i = 1, 2, \tag{12}$$

$$|| t_i ||_j = \left(\int_{\Gamma_j} t_i^2 \, d\Gamma \right)^{1/2}, \qquad i = 1, 2 \tag{13}$$

for displacements, u_i, and tractions, t_i, respectively. Equations (12) and (13) can be used directly to calculate exact boundary solution norms for each, i, solution component distribution, as well as for the resultant vector distribution.

Numerical solution norms for the p-refined BEM mesh are defined as

$$|| \hat{u}_i ||_j = \left(\int_{\Gamma_j} \hat{u}_i^2 \, d\Gamma \right)^{1/2}, \qquad i = 1, 2, \tag{14}$$

$$|| \hat{t}_i ||_j = \left(\int_{\Gamma_j} \hat{t}_i^2 \, d\Gamma \right)^{1/2}, \qquad i = 1, 2. \tag{15}$$

Equations (14) and (15) are also used to calculate component, i, and resultant solution norms for each element, j.

The solution norm ratio indicators can now be defined as

$$\theta_{u_i}^j = \frac{|| \hat{u}_i ||_j}{|| u_i ||_j}, \qquad i = 1, 2, \tag{16}$$

$$\theta_{t_i}^j = \frac{|| \hat{t}_i ||_j}{|| t_i ||_j}, \qquad i = 1, 2 \tag{17}$$

for displacement and traction, respectively, for each element, j. The solution norm ratios are calculated for component, i, and resultant vectors, and should approach 1.0 as the magnitude of the p-refined solution converges with the magnitude of the exact solution.

The relative percent errors are also used in studying local convergence for each element, j, and are defined as

$$\eta_{u_i}^{j} = \frac{||\, e_{u_i} \,||_j}{||\, u_i \,||_j} \; x \; 100\% \,, \qquad i = 1, 2 \tag{18}$$

for displacements and

$$\eta_{t_i}^{j} = \frac{||\, e_{t_i} \,||_j}{||\, t_i \,||_j} \; x \; 100\% \,, \qquad i = 1, 2 \tag{19}$$

for tractions. Relative percent errors describe the error in the p-refined solution as a percentage of the exact solution. Equations (18) and (19) are evaluated for vector component, i, as well as the resultant solution vectors. In calculating relative percent error for resultant vectors Equations (10) and (11) are substituted into Equations (18) and (19).

Global convergence of a p-refined mesh is studied by evaluating the boundary L_2 norms, solution norm ratios, and relative percent errors for the entire mesh. In addition, an equilibrium residual estimator is used to study global convergence. The L_2 error norms are evaluated for the entire mesh as

$$||\, e_{u_i} \,||_G = \left(\sum_{j=1}^{m} ||\, e_{u_i} \,||_j^2 \right)^{1/2} , \qquad i = 1, 2 \,, \tag{20}$$

$$||\, e_{t_i} \,||_G = \left(\sum_{j=1}^{m} ||\, e_{t_i} \,||_j^2 \right)^{1/2} , \qquad i = 1, 2 \tag{21}$$

where the sum on j ranges from 1 to the number of elements, m, in the mesh and $||\, e_{u_i} \,||_j$, $||\, e_{t_i} \,||_j$ are the element level norms defined in Equations (6) and (7) for each solution component, i. The error norms given in Equations (10) and (11) are used in Equations (20), and (21), respectively, for studying convergence of resultant solution vectors.

The solution norm ratios are defined on the global level as

$$\theta_{u_i}^{G} = \left(\frac{\sum_{j=1}^{m} ||\, \hat{u}_i \,||_j^2}{\sum_{j=1}^{m} ||\, u_i \,||_j^2} \right)^{1/2} , \qquad i = 1, 2 \,, \tag{22}$$

$$\theta_{t_i}^{G} = \left(\frac{\sum_{j=1}^{m} || \hat{t_i} ||_j^2}{\sum_{j=1}^{m} || t_i ||_j^2} \right)^{1/2}, \qquad i = 1, 2 \tag{23}$$

where m is the number of elements in the mesh, $|| \hat{u_i} ||_j$, $|| \hat{t_i} ||_j$ are given in Equations (14) and (15) and $|| u_i ||_j$, $|| t_i ||_j$ are given in Equations (12) and (13), respectively. The estimators in Equations (22) and (23) are evaluated for component, i, and resultant solution vectors directly.

Relative percent errors for the entire mesh are found from

$$\eta_{u_i}^{G} = \frac{|| e_{u_i} ||_G}{\left(\sum_{j=1}^{m} || u_i ||_j^2 \right)^{1/2}}, \qquad i = 1, 2 , \tag{24}$$

$$\eta_{t_i}^{G} = \frac{|| e_{t_i} ||_G}{\left(\sum_{j=1}^{m} || t_i ||_j^2 \right)^{1/2}}, \qquad i = 1, 2 \tag{25}$$

where m is the number of elements in the mesh and $|| u_i ||_j$, $|| t_i ||_j$ are given in Equations (12) and (13), respectively. Equations (24) and (25) are also evaluated for component, i, and resultant solution vectors.

Since the boundary integral formulation for elastostatic problems is based on equilibrium of forces in the global x and y directions, one can define the equilibrium residual, E_R, as

$$E_R = \sqrt{\hat{F_1}^2 + \hat{F_2}^2} \tag{26}$$

where

$$\hat{F_1} = \sum_{j=1}^{m} \int_{\Gamma_j} \hat{t_1} \, d\Gamma , \tag{27}$$

$$\hat{F_2} = \sum_{j=1}^{m} \int_{\Gamma_j} \hat{t_2} \, d\Gamma . \tag{28}$$

Traction distributions $\hat{t}_1$, $\hat{t}_2$ in Equations (27) and (28), respectively, are the p-refined solutions obtained for each element, j, from Equation (3). The equilibrium residual, E_R, is a measure of how well static equilibrium is satisfied during refinement and is equal to 0.0 for the exact solution.

5. Demonstration Problem - L-shaped domain

The L-shaped domain is obtained from taking a quarter symmetry model of a square plate, with a square hole, loaded in tension. The geometry and boundary conditions for the quarter symmetry model are shown in Figure 1. The L-shaped problem is of interest because of the geometric singularities which give rise to discontinuous traction distributions and a stress concentration at point C in Figure 1. This problem was studied by Parreira [14] and Cerrolaza et al. [7] for convergence of p-adaptive, BEM algorithms and by Zienkiewicz et al. [24] for convergence of the p-adaptive FEM. Both of the BEM works [7, 14] used the Legendre family of polynomials as refinement basis functions. Zienkiewicz [24] points out that h-convergence for the FEM should be of order $O(h^2)$ when using an energy error norm. This translates to a 1:1 Log-Log convergence slope when plotting energy error versus degrees of freedom.

The initial discretization for the L-shaped plate is defined by mesh points A through F, as shown in Figure 1. An initial mesh of 6 elements is used in both the h- and p-refinement analyses. The subsequent h-refined

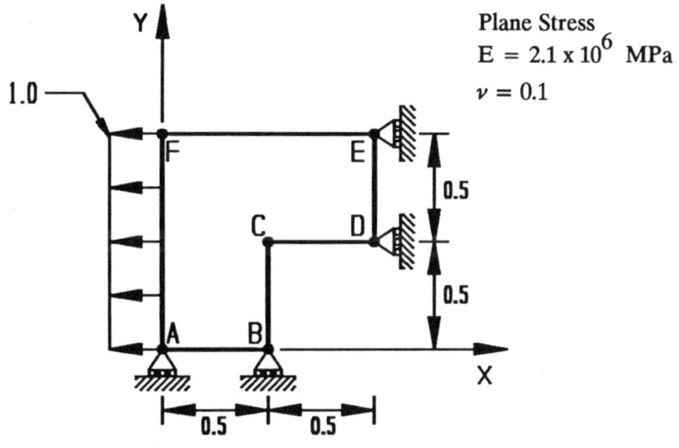

Figure 1. L-shape domain problem and initial mesh.

meshes are generated using uniform, linear elements of length 1/2 m, 1/6 m, and 1/10 m. During p-refinement the initial mesh size remains constant while

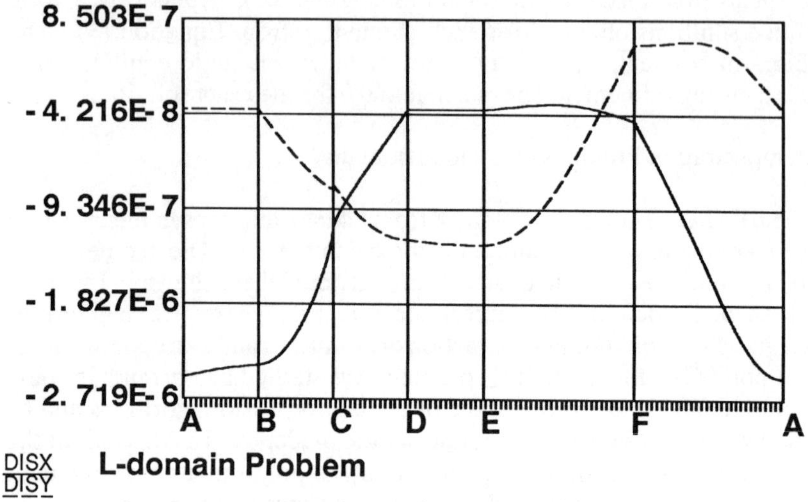

Note: Units for the ordinate and abscissa are m.

Figure 2. Displacements from 120 element mesh.

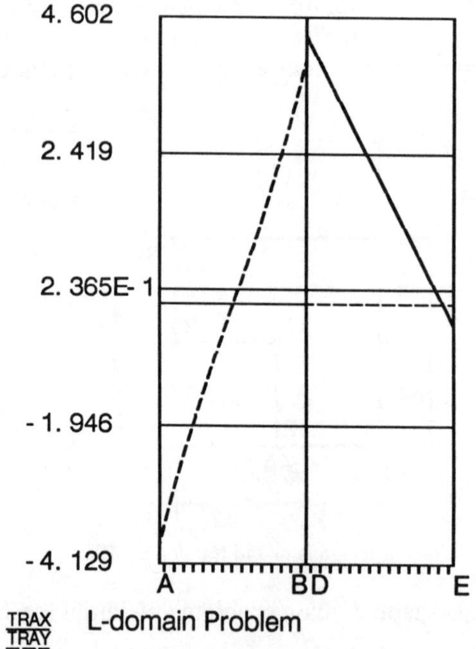

Note: Units for the ordinate are m and for the abscissa MPa.

Figure 3. Tractions from 120 element mesh.

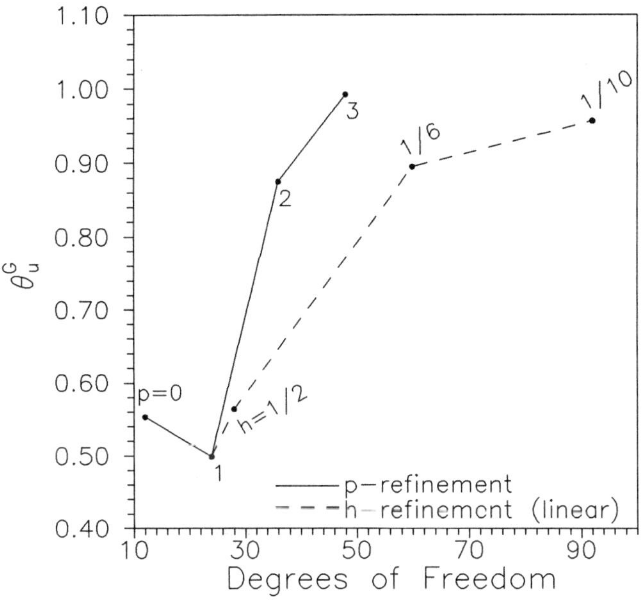

Figure 4. Global solution norm ratio for displacement.

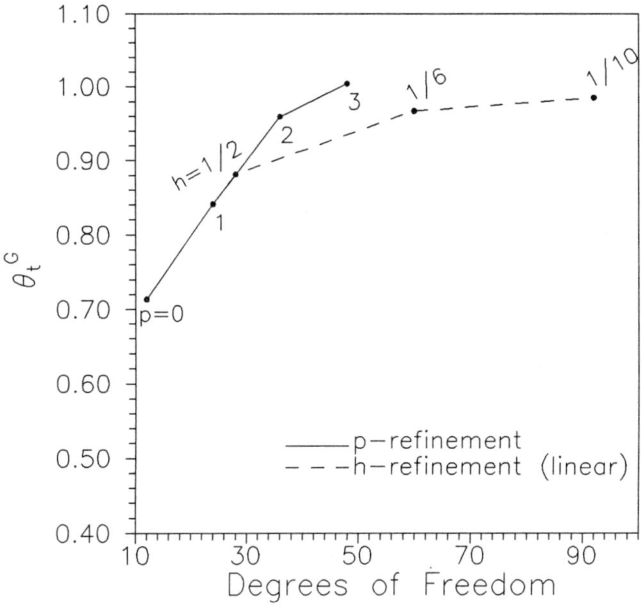

Figure 5. Global solution norm ratio for traction.

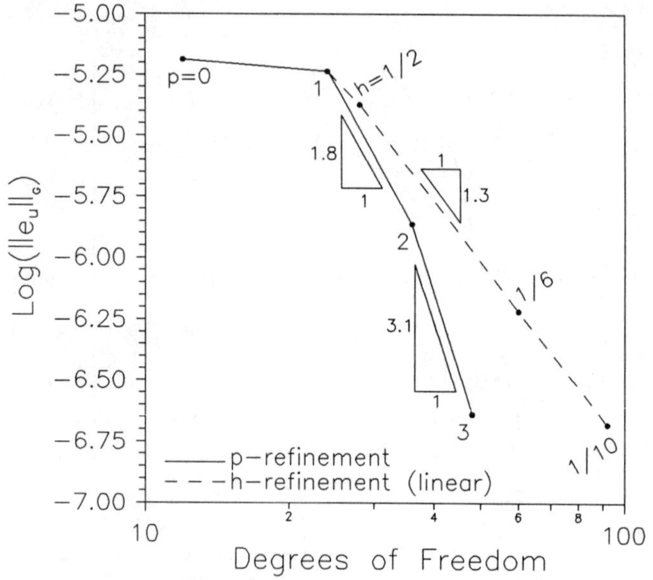

Figure 6. Global boundary error norm for displacement.

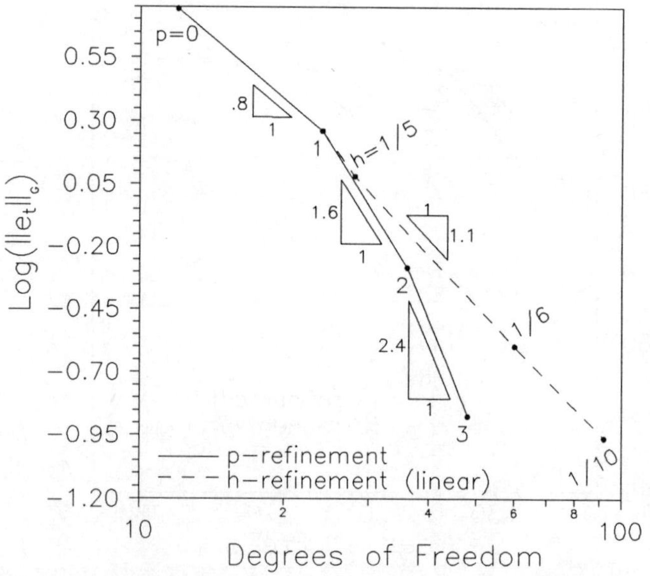

Figure 7. Global boundary error norm for traction.

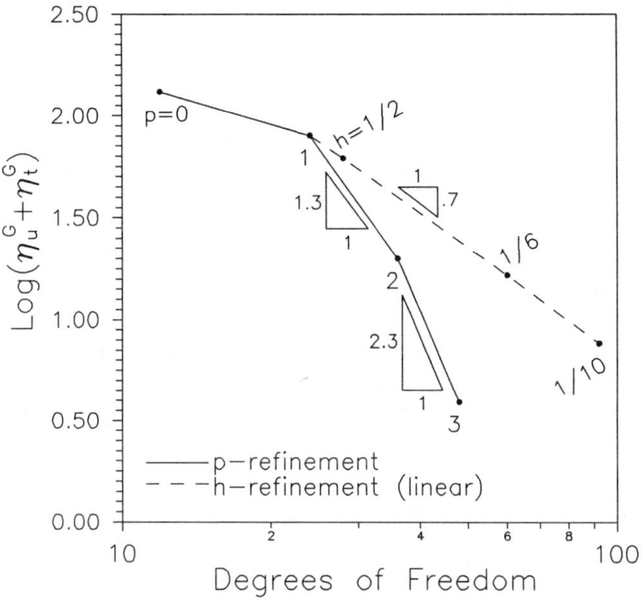

Figure 8. Sum of global relative percent errors.

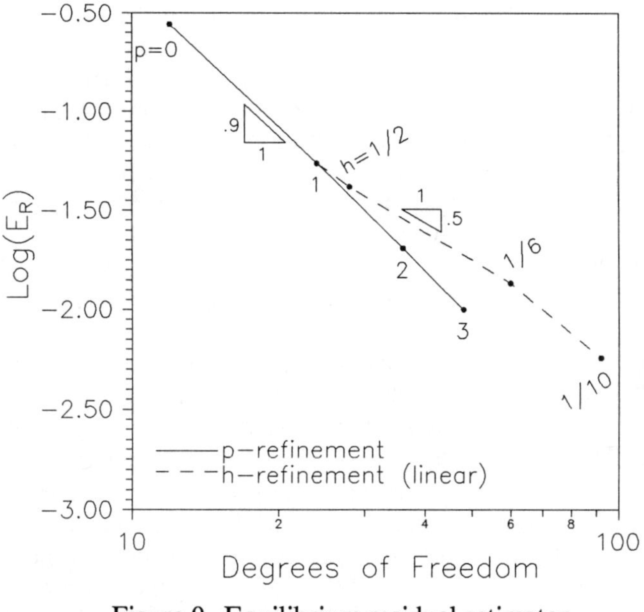

Figure 9. Equilibrium residual estimator.

the order of the solution is varied as constant, linear, quadratic, and cubic distributions.

Global convergence is studied using the error estimates defined in Section 4. The reference for convergence is taken to be the solution obtained from BEASY using a uniform, linear mesh of 120 quadratic elements. Figures 2 and 3 show the displacement and traction distributions, respectively, over the boundary of the L-shaped domain. Only the tractions over sides AB and DE are shown in Figure 3 because traction is prescribed as a boundary condition on all other sides.

Figures 4 and 5 show convergence of the global solution norm ratios for displacement and traction, respectively, as a function of the number of degrees of freedom. As expected, the slope of the solution norm ratio for displacement is steeper than the slope of the solution norm ratio for traction. This is because the displacement distribution shown in Figure 2 is better suited to higher-order interpolation whereas the nearly linear traction distributions shown in Figure 3 are less effected by the higher-order terms. In addition, the traction solution norm ratio starts closer to 1.0 than the solution norm ratio for displacement because of the nearly linear traction distributions shown in Figure 3. Note that during p-refinement both displacement and traction converge closer to the exact solution, faster than h-refinement using linear, continuous interpolation.

Figures 6 and 7 show Log-Log convergence of the global L_2 boundary error norms for displacement and traction, respectively. One should note that the constant element, p-refined mesh in Figure 6 does not accurately capture the higher-order displacement distribution as is further suggested in Figure 4. The p-convergence between linear and quadratic meshes is, however, 1.4 times faster than h-convergence for both displacement and traction. In going from quadratic to cubic interpolation p-refinement is 2.4 times faster than h-convergence for displacement and 2.2 times faster for traction, respectively. The traction and displacement slopes for h-convergence are slightly greater than the asymptotic, 1:1 ratio stated in [24], because of the differences between the L_2 and energy error norms as explained in [15].

Figures 8 and 9 show global convergence of the relative percent error and equilibrium residual. In Figure 8 the Log of the sum of the global relative percent errors (η_u, η_t) is plotted versus the Log of the corresponding degrees of freedom. The convergence slope of 0.7 for h-refinement and 1.3 for linear to quadratic p-refinement, is in agreement with the study done by Cerrolaza et al. in [7]. The p-convergence slope of 2.3 between quadratic and cubic interpolation, however, is not in agreement, surpassing even the 2.0 slope obtained for p-adaptive refinement [7]. This discrepancy is probably due to the relative p-refinement strategy used in [7] which employs Legendre polynomials. Figure 9 shows a Log-Log plot of the equilibrium residual, E_R, verses degrees of freedom. Here convergence for the p-refined mesh is approximately 1.8 times faster than the h-refined mesh. In all of the global convergence plots shown in Figures 4 through 9, the absolute, p-refined mesh approaches the fine mesh solution faster than the h-refined mesh.

5. CONCLUSIONS

A 2-D, BE elasticity program was developed to study absolute p-convergence of discontinuous constant, and continuous linear through cubic solution distributions. Absolute p-refinement eliminates the question of where to collocate for hierarchial degrees of freedom associated with relative p-refinement schemes. The proposed p-refinement strategy is compared to linear, h-refinement through the analysis of a tensile plate with a square hole. A convergence study shows that the p-refined mesh approaches the exact solution considerably faster than uniform h-refinement for the demonstration problem. The absolute p-refinement strategy is shown to be an effective means of reducing numerical error, especially during higher-order refinement.

6. ACKNOWLEDGEMENT

The authors would like to thank Professor Kenneth E. Scott, Acting Head of the Mechanical Engineering Department at Worcester Polytechnic Institute for providing the publication support for this work. In addition, the use of the Computational Mechanics Laboratory in the Mechanical Engineering Department at WPI is gratefully appreciated.

7. REFERENCES

1. Alarcón, E., Reverter, A., and Molina, J. Advances in Finite Element Technology - Hierarchial Boundary Elements, Computers and Structures, Vol. 20, pp.151-156, 1985.

2. Alarcón, E., and Reverter, A., p-Adaptive Boundary Elements, International Journal for Numerical Methods in Engineering, Vol. 23, pp. 801-829, 1986.

3. Banerjee, P. K., and Butterfield, R., Two-Dimensional Problems of Elastostatics. Chapter 4, Boundary Element Methods in Engineering Science, pp. 177-215, McGraw-Hill, London, 1981.

4. Banerjee, P. K., and Butterfield, R., Parametric Representation of Functions and Geometry. Chapter 8, Boundary Element Methods in Engineering Science, pp. 177-215, McGraw-Hill, London, 1981.

5. BEASY, Users Manual, Version 3.0, Computational Mechanics Publications, Ashurst, Southampton, U. K., 1986.

6. Brebbia, C. A., Telles, J. C. F., and Wrobel, L. C., Boundary Element Techniques, Springer-Verlag, Berlin and New York, 1984.

7. Cerrolaza, M. and Alarcón, E., Further Applications of p-Adaptive Boundary Elements, Boundary Elements X, (Ed. Brebbia, C. A.), Computational Mechanics Publications, Southampton, 1988.

8. Cerrolaza, M. and Alarcón, E., p-Adaptive Boundary Elements for Three-Dimensional Potential Problems, Communications in Applied Numerical Methods, Vol. 3, pp. 335-344, 1987.

9. FIESTA: Theoretical Literature, Instituto Sperimentale Modelli Strutture (ISMES), Bergamo, Italy, 1984.

10. Kardestuncer, H. and Norrie, D. H., Part 2 - Other Finite-Element Methods. Chapter 6, Finite Element Handbook, McGraw-Hill, New York, 1987.

11. Lera, S. Gómez, Cerrolaza, M., and Alarcón, E., Adaptive Refinements in BEM, Boundary Elements IX, (Eds. Brebbia, C. A., Wendland, W. L., and Kuhn, G.), Computational Mechanics Publications, Southampton, 1987.

12. Mitra, A. K. and Ingber, M. S., Resolving Difficulties in the BIEM Caused by Geometric Corners and Discontinuous Boundary Conditions, Boundary Elements IX, (Eds. Brebbia, C. A., Wendland, W. L., Kuhn, G.), Computational Mechanics Publications, Southampton, 1987.

13. Shephard, M. S., Approaches to the Automatic Generation and Control of Finite Element Meshes, Applied Mechanics Review, Vol. 41, No. 4, pp. 169-185, 1988.

14. Parreira, P., Self-Adaptive p-Hierarchical Boundary Elements in Elastostatics, Boundary Elements IX, (Eds. Brebbia, C. A., Wendland, W. L., Kuhn, G.), Computational Mechanics Publications, Southampton, 1987.

15. Reddy, J. N., Finite Element Analysis of One-Dimensional Problems. Chapter 3, An Introduction to the finite Element Method, McGraw-Hill, New York, 1984.

16. Rencis, J. J., Urekew, T. J., Jong, K-Y., Kirk, R. and Federico, P., *A Posteriori* Error Estimator for Finite Element and Boundary Element Methods, Journal of Applied Mathematical Modeling, (submitted for review).

17. Reverter, A., Gónzalez, A.and Alarcón, E., Indicators and Estimators in p-Adaptive Boundary Elements, Boundary Elements VII, (Eds. Brebbia, C. A. and Maier, G.), Computational Mechanics Publications, Southampton, 1985.

18. Riccardella, P. C., An Improved Implementation of the Boundary-Integral Technique for Two Dimensional Elasticity Problems, Report SM- 72-86, Department of Mechanical Engineering, Carnegie-Mellon University, 1972.

19. Noor, A. K. and Babuška, I., Quality Assessment and Control of Finite Element Solutions, Finite Elements in Analysis and Design, 3, pp. 1-26, 1987.

20. Szabó, B. A., PROBE: Theoretical Manual, NOETIC Technology Corp., 7980 Clayton Road, Suite 205, St. Louis, MO. 63117.

21. Szabó, B. A. and Peano, A., COMET-PR: The First Computer Implementation of the P-Version of the Finite Element Method, Finite Element Systems - A Handbook, (Ed. Brebbia, C. A.), pp. 147-158, Springer-Verlag, Berlin, 1982.

22. Umetani, S-I., Adaptive Boundary Element Methods in Elastostatics, Computational Mechanics Publications, Southampton, 1988.

23. Wiberg, N-E., and Möller, P., Formulation and Solution of Hierarchial Finite Element Equations, International Journal for Numerical Methods in Engineering, Vol. 26, pp. 1213-1233, 1988.

24. Zienkiewicz, O. C., Gago, J., and Kelly, D. W., The Hierarchical Concept in Finite Element Analysis, Computers and Structures, Vol. 16, pp. 53-65, 1983.

An h-p Version of BEM for Plane Mixed Boundary Value Problems

B.Q. Guo*, T. von Petersdorff** and E.P. Stephan***

**Engineering Mechanics Research Corporation Troy, Michigan, USA*

***Fachberich Mathematik, TH Dartmstadt, FRG*

****School of Mathematics, Georgia Institute of Technology, Atlanta, GA, USA*

ABSTRACT

We present the h-p version of the boundary element method for mixed boundary value problems on plane domains with piecewise analytic boundary and data. The exponential rate of convergence of the boundary element Galerkin solution is obtained when a geometric mesh refinement is used near the vertices, and near the collision points where the boundary conditions change.

INTRODUCTION

Let $\Omega \subset \mathbb{R}^2$ be a bounded domain with polygonal boundary $\Gamma = \bar{\Gamma}_1 \cup \bar{\Gamma}_2$ consisting of straight sides Γ^j with endpoints A_j, A_{j+1}, $j = 1,\dots,M$ where $A_{M+1} = A_1$. We consider the mixed boundary value problem of the Laplacian

$$\begin{aligned} \Delta u &= 0 \text{ in } \Omega \\ u &= g_1 \text{ on } \Gamma_1 \\ \frac{\partial u}{\partial n} &= g_2 \text{ on } \Gamma_2 \end{aligned} \qquad (1)$$

For $g_1 \in H^{1/2}(\Gamma_1)$, $g_2 \in H^{-1/2}(\Gamma_2)$ problem (1) has due to Costabel and Stephan [1] a unique solution $u \in H^1(\Omega)$. For the definition of the Sobolev spaces $H^1(\Omega)$, $H^{1/2}(\Gamma_1)$, $H^{-1/2}(\Gamma_2)$ see [2], [3], [1].

In the following we solve problem (1) via an h-p version of the boundary element Galerkin method. First we convert problem (1) in an equivalent system of strongly elliptic boundary integral equations on Γ.

For $x \in \Gamma_k$ and sufficiently smooth f we introduce the integral operators

$$V_{jk}f(x) := -\frac{1}{\pi}\int_{\Gamma_j} \ell n|x-y|f(y)ds_y$$

$$K_{jk}f(x) := -\frac{1}{\pi}\int_{\Gamma_j} \frac{\partial}{\partial n_y}\{\ell n|x-y|\}f(y)ds_y$$

$$K'_{jk}f(x) := -\frac{1}{\pi}\int_{\Gamma_j} f(y)\frac{\partial}{\partial n_x}\{\ell n|x-y|\}ds_y$$

$$D_{jk}f(x) := \frac{1}{\pi}\frac{\partial}{\partial n_x}\int_{\Gamma_j} \frac{\partial}{\partial n_y}\{\ell n|x-y|\}f(y)ds_y.$$

These are the operators of the single layer potential V_{jk}, of the double potential K_{jk}, its adjoint K'_{jk} and the operator D_{jk} of the normal derivative of K_{jk}. Let $\ell,\tilde{\ell}$ denote arbitrary extensions from $H^{1/2}(\Gamma_1)$ to $H^{1/2}(\Gamma)$ and from $H^{-1/2}(\Gamma_2)$ to $H^{-1/2}(\Gamma)$, respectively, and let $f|_{\Gamma_k}$ denote the restriction of f to Γ_k. Costabel and Stephan [1] showed the following result for the system of integral equations corresponding to (1). Here we need for $s \in \mathbb{R}$ the space

$$\tilde{H}^s(\Gamma_k) = \{f \in H^s(\Gamma): \text{supp } f \subset \bar{\Gamma}_k\}$$

Theorem 1 [1]: For $g_1 \in H^{1/2}(\Gamma_1)$ and $g_2 \in H^{-1/2}(\Gamma_2)$ the system of boundary integral equations

$$\mathcal{A}\binom{v^*}{\psi^*} = \mathcal{B}\binom{g_1}{g_2} - \mathcal{A}\binom{\ell g_1|_{\Gamma_2}}{\tilde{\ell} g_2|_{\Gamma_1}} \tag{2}$$

with the boundary integral operators

$$\mathcal{A} := \begin{bmatrix} D_{22} & K'_{12} \\ -K_{21} & V_{11} \end{bmatrix}, \quad \mathcal{B} := \begin{bmatrix} -D_{12} & 1-K'_{22} \\ 1+K_{11} & -V_{21} \end{bmatrix}$$

has a unique solution $(v^*,\psi^*) \in \tilde{H}^{1/2}(\Gamma_2) \times \tilde{H}^{-1/2}(\Gamma_1)$. Problems (1) and (2) are equivalent in the sense that (v^*,ψ^*) yields the unknown boundary data $u|_{\Gamma_2}$, $\frac{\partial u}{\partial n}\Big|_{\Gamma_1}$ of the solution $u \in H^1(\Omega)$ of (1) by

$$u|_{\Gamma_2} = v^* + \ell g_1|_{\Gamma_2}, \quad \frac{\partial u}{\partial n}\Big|_{\Gamma_1} = \psi^* + \tilde{\ell} g_2|_{\Gamma_1} \tag{3}$$

Conversely, let $u \in H^1(\Omega)$ solve (1) then (v^*,ψ^*) obtained from (3) solve (2).

THE h-p VERSION WITH GEOMETRIC MESH

Next we consider the Galerkin method for the system of integral equations (2). By introducing a geometric mesh refinement towards each vertex of Γ, we define the boundary element spaces needed for the h-p version as follows. Let d_j denote the length of the side Γ^j. First bisect each side Γ^j into two parts Γ^j_- and Γ^j_+ such that $A_j \in \overline{\Gamma^j_-}$ and $A_{j+1} \in \overline{\Gamma^j_+}$. Then choose a mesh parameter $0 < \sigma < 1$ and an integer $n \geq 0$, and introduce the meshpoints $A^{j,k}_-$ on Γ^j_- with $\mathrm{dist}(A^{j,k}_-, A_j) = \sigma^{n-k}\frac{d_j}{2}$ for $k = 0,\ldots,n$. This defines n subintervals $\Gamma^{j,k}_-$ with endpoints $A^{j,k-1}_-$, $A^{j,k}_-$ for $k = 1,\ldots,n$. Define analgously the subintervals $\Gamma^{j,k}_+$ with endpoints $A^{j,k-1}_+$, $A^{j,k}_+$ on Γ^j_+. Then on this geometric mesh we introduce the boundary element spaces

$$\begin{aligned} S_{n,2} &:= \{v \in C^0(\Gamma_2)\colon v|_{\Gamma^{j,k}_{(\pm)}} \in P_k(\Gamma^{j,k}_{(\pm)}) \text{ for } \Gamma^{j,k}_{(\pm)} \subset \Gamma^j \subset \Gamma_2,\ k = 1,\ldots,n\} \\ S_{n,1} &:= \{\psi \in L^2(\Gamma_1)\colon \psi|_{\Gamma^{j,k}_{(\pm)}} \in P_k(\Gamma^{j,k}_{(\pm)}) \text{ for } \Gamma^{j,k}_{(\pm)} \subset \Gamma^j \subset \Gamma_1,\ k = 1,\ldots,n\} \end{aligned} \tag{4}$$

where $P_k(\Gamma^{j,k}_{(\pm)})$ denotes the space of polynomials of degree $\leq k$ on $\Gamma^{j,k}_{(\pm)}$.

Then the *Galerkin method* for problem (2) reads as: *For given* $g_1 \in H^{1/2}(\Gamma_1)$ *and* $g_2 \in H^{-1/2}(\Gamma_2)$ *find* $v^*_n \in S_{n,2}$, $\psi^*_n \in S_{n,1}$ *satisfying for all* $w_n \in S_{n,2}$, $\phi_n \in S_{n,1}$

$$\left\langle \mathcal{A}\binom{v^*_n}{\psi^*_n}, \binom{w_n}{\phi_n} \right\rangle = \left\langle \mathcal{B}\binom{g_1}{g_2} - \mathcal{A}\binom{\ell g_1|_{\Gamma_2}}{\tilde{\ell} g_2|_{\Gamma_1}}, \binom{w_n}{\phi_n} \right\rangle \tag{5}$$

where $\langle\cdot,\cdot\rangle$ *denotes the* L^2*-scalar product.*

Here ℓg_1 and $\tilde{\ell} g_2$ denote the even and the odd extensions to Γ of g_1 and g_2, respectively (for details see Guo, von Petersdorff, Stephan [4]).

In [4] we show exponential convergence for the Galerkin method (5) if the data g_1, g_2 are piecewise analytic. Then g_1, g_2 belong to the countable normed spaces $B^{1,2}_{\hat{\beta}}(\Gamma_1)$ and $B^{0,1}_{\hat{\beta}}(\Gamma_2)$, respectively. (For the definition of those spaces see Babuška, Guo [5].)

Theorem 2 [4]: *Let* $g_1 \in C^0(\Gamma_1) \cap B^{1,2}_{\hat{\beta}}(\Gamma_1)$ and $g_2 \in B^{0,1}_{\hat{\beta}}(\Gamma_2)$ *be given. Then for sufficiently large* n *the Galerkin equations* (5) *have a unique solution* $v^*_n \in S_{n,2}$, $\psi^*_n \in S_{n,1}$. *Furthermore with the exact solution* $v^* \in \tilde{H}^{1/2}(\Gamma_2)$, $\psi^* \in \tilde{H}^{-1/2}(\Gamma_2)$ *of*

(2) *there holds*

$$||v^* - v_n^*||_{\tilde{H}^{1/2}(\Gamma_2)} + ||\psi^* - \psi_n^*||_{\tilde{H}^{-1/2}(\Gamma)} \leq Ce^{-b\sqrt{N}} \tag{6}$$

where $N := \dim S_{n,2} + \dim S_{n,1}$ *and* $C, b > 0$ *are constants independent of* n.

Corollary 1 [4]: *Under the hypothesis of* Theorem 2 *there holds with the exact solution* $u \in H^1(\Omega)$ *of problem* (1)

$$||u|_{\Gamma_2} - v_n||_{\tilde{H}^{1/2}(\Gamma_2)} + \left\|\frac{\partial u}{\partial n}\Big|_{\Gamma_1} - \psi_h\right\|_{\tilde{H}^{-1/2}(\Gamma_1)} \leq Ce^{-b\sqrt{N}}$$

where C *and* $b > 0$ *are independent of* N *and*

$$v_n := \ell g_1|_{\Gamma_2} + v_n^*, \quad \psi_n := \tilde{\ell} g_2|_{\Gamma_1} + \psi_n^*$$

are given via the Galerkin solutions $v_n^* \in S_{n,2}$, $\psi_n^* \in S_{n,1}$ *of* (5).

For the pure Dirichlet and the pure Neumann problems of the Laplacian, i.e., (1) with $\Gamma_2 = \emptyset$ and $\Gamma_1 = \emptyset$, respectively, the exponential convergence of the corresponding h-p version of the Galerkin scheme (5) was already shown in Babuška, Guo, Stephan [6]. For crack problems

$$\Delta u = 0 \text{ in } \mathbb{R}^2 \backslash \bar{\Gamma} \text{ with } u = O(\frac{1}{|x|}) \text{ as } |x| \to \infty \tag{7}$$

$$u = g_1 \text{ on } \Gamma \quad \text{or} \quad \frac{\partial u}{\partial n} = g_2 \text{ on } \Gamma$$

with Γ being a smooth, open arc, the solutions of the corresponding integral equations

$$V\psi = g_1 \text{ on } \Gamma \quad \text{or} \quad Dv = g_2 \text{ on } \Gamma \tag{8}$$

satisfy $\psi = \left[\frac{\partial u}{\partial n}\right]_\Gamma \in \tilde{H}^{-1/2}(\Gamma)$ and $v = [u]|_\Gamma \in \tilde{H}^{1/2}(\Gamma)$ (see Stephan, Wendland [7, 8]). These solutions show near the crack tip the same behavior as the solution (v^*, ψ^*) of (2) at the collision points in $\bar{\Gamma}_1 \cap \bar{\Gamma}_2$. This can be seen by extending the crack Γ to a smooth, closed curve $\tilde{\Gamma}$. Then we observe for the solution $u \in H^1_{loc}(\mathbb{R}^2\backslash\bar{\Gamma})$ of the Dirichlet problem (7) that its normal derivative $g_2 := \frac{\partial u}{\partial n}$ on $\tilde{\Gamma}\backslash\bar{\Gamma}$ is smooth. Hence this solution u solves also the mixed boundary problem (1) with $u = g_1$ on Γ and $\frac{\partial u}{\partial n} = g_2$ on $\tilde{\Gamma}\backslash\bar{\Gamma}$. Correspondingly, one can convert the Neumann problem (7) into a mixed boundary value problem of form (1). Therefore for the Galerkin scheme corresponding to the integral equations in (8) there hold estimates of the form (6) guaranteeing the exponential convergence of the Galerkin solutions if a geometric mesh refinement towards the crack tips and boundary element spaces as in (4) are used (see Stephan [9]).

NUMERICAL RESULTS

In the following we present numerical experiments for the Dirichlet and Neumann problems (7) underlying the convergence estimate (6).

Example 1: We consider

$$V\psi = 2x \text{ on } \Gamma = [-1,1] \tag{9}$$

with the true solution $\psi = \frac{-2x}{\sqrt{1-x^2}}$. Tables 1, 2 show the relative error of the Galerkin solution $\psi_n \in S^0_{n,1}$ in the energy norm $||\psi-\psi_n||_V$ vs. degrees of freedom N. Note $\psi_n \in S^0_{n,1}$ satisfies

$$\left\langle V\psi_n,\phi_n\right\rangle = \left\langle 2x,\phi_n\right\rangle \tag{10}$$

for all $\phi_n \in S^0_{n,1}$ where ϕ_n is a piecewise polynomial of degree $k-1$ on the subinterval $\Gamma^{1,k}_{(\pm)}$, $k = 1,...,n$, where $\Gamma^{1,k}_{(\pm)}$ is defined as above with $\Gamma = \Gamma^1$. Here we set

$$||\psi||^2_V := \left\langle V\psi,\psi\right\rangle \simeq ||\psi||^2_{\tilde{H}^{-1/2}(\Gamma)} \tag{11}$$

yielding $||\psi||_V = \sqrt{2}\pi$. Here the estimate (6) gives

$$||\psi-\psi_n||^2_V \leq \tilde{C}e^{-\tilde{b}\sqrt{N}} \tag{12}$$

where $\tilde{C} = C^2$, $\tilde{b} = 2b$ with C,b as in (6). Note N coincides with the dimension of the boundary element space $S^0_{n,1}$. If we assume that equality holds in (12) for two consecutive values of n, then we obtain two equations which allow to compute numerical values $\tilde{b}$,$\tilde{C}$ for each n > 1. These values are listed in Tables 1 and 2. Since these numbers appear to converge they confirm our theoretical convergence estimate (6). The speed of convergence (measured by b) depends on the mesh parameter σ. Our numerical experiments indicate that the optimal value of σ is 0.1. Tables 1 and 2 show slower convergence for $\sigma = 0.3$, but the values of $\tilde{C}$ and $\tilde{b}$ stabilize more quickly in this case.

Table 1. Exponential Convergence of the h-p Method for Scheme (10) with $\sigma = 0.1$

n	$N = \dim S^0_{n,1}$	$\frac{\|\psi-\psi_n\|_V}{\|\psi\|_V}$	$\tilde{b}$	$\tilde{C}$
1	2	52.8%		
2	6	15.6%	2.359	154.5
3	12	5.2%	2.166	96.3
4	20	1.7%	2.176	100.1

Table 2. Exponential Convergence of the h-p Method for Scheme (10) with $\sigma = 0.3$

n	$N = \dim S^0_{n,1}$	$\frac{\|\psi-\psi_n\|_V}{\|\psi\|_V}$	$\tilde{b}$	$\tilde{C}$
1	2	52.8%		
2	6	27.2%	1.278	33.51
3	12	14.5%	1.237	30.28
4	20	7.9%	1.208	27.47
5	30	4.3%	1.202	26.68

Example 2: We consider

$$Dv = 1 \text{ on } \Gamma = [-1,1] \tag{13}$$

with the true solution $v = 2\sqrt{1-x^2}$. Tables 3 and 4 show the relative error in the energy norm $\|v-v_n\|_D$ of the Galerkin solution $v_n \in S_{n,2}$ vs. degress of freedom N.

Note $v_n \in S_{n,2}$ satisfies

$$\left\langle Dv_n, w_n \right\rangle = \left\langle 1, w_n \right\rangle \tag{14}$$

for all $w_n \in S_{n,2}$ where w_n is a continuous, piecewise polynomial of degree k on $\Gamma^{1,k}_{(\pm)}$, $k = 1,\ldots,n$. Here we set (where $\simeq$ denotes equivalence of norms)

$$||v||^2_D := \left\langle Dv, v \right\rangle \simeq ||v||^2_{\tilde{H}^{1/2}} \tag{15}$$

yielding $||v||_D = \sqrt{\pi}$. Now the estimate (6) gives

$$||v - v_n||^2_D \leq \tilde{C} e^{-\tilde{b}\sqrt{N}} \tag{16}$$

where $\tilde{C} = C^2$, $\tilde{b} = 2b$ with C,b as in (6). Now N coincides with the dimension of the boundary element space $S_{n,2}$. The numerical values $\tilde{C}, \tilde{b}$ (listed in Tables 3 and 4) seem to converge and thus confirm our theoretical convergence estimate (6). Again the speed of convergence of the relative error in the energy norm depends on the mesh parameter σ.

Table 3. Exponential Convergence of the h-p Method for Scheme (14) with $\sigma = 0.3$

n	$N = \dim S_{n,2}$	$\frac{\lVert v-v_n \rVert_D}{\lVert v \rVert_D}$	$\tilde{b}$	$\tilde{C}$
1	1	52.8%		
2	5	18.9%	1.659	4.60
3	11	7.3%	1.764	5.82
4	19	2.8%	1.817	6.94
5	29	1.1%	1.840	7.67

Table 4. Exponential Convergence of the h-p Method for Scheme (14) with $\sigma = 0.5$

n	$N = \dim S_{n,2}$	$\frac{\|\|v-v_n\|\|_D}{\|\|v\|\|_D}$	$\tilde{b}$	$\tilde{C}$
1	2	52.8%		
2	5	35.9%	.623	1.63
3	11	24.6%	.701	1.94
4	19	17.1%	.693	1.89
5	29	12.0%	.689	1.86

CONCLUSION

The h-p method on a geometric mesh gives exponential convergence, therefore it is superior to the pure h- and p-methods which give only algebraic convergence rates: For an h-method with fixed degree p one obtains for (8) on an optimally graded mesh (see von Petersdorff [10] for p = 0)

$$||\psi-\psi_h||_{\tilde{H}^{1/2}(\Gamma)} \leq Ch^{(p+3)/2} \tag{17}$$

Here the number of degrees of freedom is proportional to h^{-1}. For a p-method with a fixed number of intervals we obtain for problem (8) with arbitrary $\epsilon > 0$

$$||\psi-\psi_p||_{\tilde{H}^{-1/2}(\Gamma)} \leq Cp^{1-\epsilon} \tag{18}$$

where the number of degrees of freedom is proportional to p (see Stephan, Suri [11]). Similar results hold for arbitrary mixed boundary value problems.

ACKNOWLEDGEMENT

The third author was partially supported by the NSF Grant DMS-8704463.

REFERENCES

[1] Costabel, M. Stephan, E. P.: Boundary Integral Equations for Mixed Boundary Value Problems in Polygonal Domains and Galerkin Approximation, Mathematical Models and Methods in Mechanics, Banach Center Publications 15, Warsaw, 175-251, 1985.

[2] Lions, J. L., Magenes, D.: Non-Homogeneous Boundary Value Problems and Applications I, Springer, Berlin, 1972.

[3] Grisvard, P.: Elliptic Problems in Nonsmooth Domains, Pitman, Boston, 1985.

[4] Guo, B. Q., von Petersdorff, T., Stephan, E. P.: On the Exponential Convergence of the h-p Boundary Element Galerkin Method for Mixed Boundary Value Problems on Polygonal Domains, to appear.

[5] Babuška, I., Guo, B. Q.: Regularity of the Solution of Elliptic Problems with Piecewise Analytic Data. Part 2: The Trace Spaces and Application to Boundary Value Problems with Non-Homogeneous Conditions, to appear in SIAM J. Math. Anal.

[6] Babuška, I., Guo, B. Q., Stephan, E. P.: On the Exponential Convergence of the h-p Version of the Boundary Element Galerkin Method on Polygons, to appear.

[7] Stephan, E. P., Wendland, W. L.: An Augmented Galerkin Procedure for the Boundary Integral Method Applied to Two-Dimensional Screen and Crack Problems, Applicable Analysis 18, 183-219, 1984.

[8] Stephan, E. P., Wendland, W. L.: A Hypersingular Boundary Integral Method for Two-Dimensional Screen and Crack Problems, Archive of Rational Mechanics and Analysis, to appear.

[9] Stephan, E. P.: The h-p Version of the Galerkin Boundary Element Method for Integral Equations on Polygons and Open Arcs. In Proc. Conference BEM-10, Southhampton (ed. C. A. Brebbia) 1988.

[10] von Petersdorff, Tobias: Elasticity Problems in Polyhedra – Singularities and Approximations with Boundary Element Methods. TH Darmstadt, Thesis 1989.

[11] Stephan, E. P., Suri, M.: On the Convergence of the p-Version of the Boundary Element Galerkin Method, Mathematics of Computation 52, 31-48, 1989.

Further Developments on Error Indicators and Estimators for Adaptive Hierarchical Boundary Elements

Pedro Parreira

Department of Civil Engineering, Instituto Superior Técnico, Av. Rovisco Pais, 1096 Lisboa, Portugal

SUMMARY

The use of the Legendre polynomials or the Peano's family in p-hierarchic adaptive refinements in the boundary element method (BEM) is discussed. The collocation and the Galerkin approaches, when referred to such refinements, are analysed. A mixed collocation-Galerkin method is proposed.

An indirect method to evaluate the integral of the residue over the boundary, required to compute error indicators and estimators, is presented. Its implementation on self-adaptive BEM codes is discussed and it is tested in classic plane stress problems.

INTRODUCTION

The success of an adaptive process is highly dependent on the definition of error indicators, which point out "where" additional refinements are more effective, and error estimators, which give a "measure of the error" to determine whether further refinements are necessary[1].

In the BEM, hierarchical adaptive stratagies have been studied by different researchers since 1984[2,3]. The p-version[4,5,6,7,8,9,10,11,12], the h-version[3,9] and recently the hp-version[13] of adaptive BEM refinements have been presented using either collocation[4,5,6,8,11,12,13] or Galerkin[3,7,9,10] procedures. In these works error indicators and estimators are defined through a "residual displacement" boundary function, which is obtained from the Somigliana's identity by introducing the discrete boundary element solution. However, a direct evaluation of this function involves a very high computational effort, which may discourage the use of adaptive software in industrial applications.

In this paper, the use of the Legendre polynomials or Peano's family in p-hierarchic adaptive BEM is discussed. Error indicators and estimators are obtained through the residue, either in a H^0 (or H^1) norm or in an "energy norm". The direct computation of the residue is analysed. The use of collocation or Galerkin methods on adaptive BEM is discussed and a mixed collocation-Galerkin approach is proposed. An indirect method for the evaluation of the residue is presented. This new method, which is based on a series expansion of the residue in each element, has the advantage of not requiring a substantial computational effort when implemented in a self-adaptive hierarchical code.

BOUNDARY ELEMENT DISCRETIZATION

A broad class of engineering problems formulated by the direct boundary element approach are based on the boundary integral equation[14]

$$c_{ij}(\mathbf{z})\,u_j(\mathbf{z}) + \fint_\Gamma t^\star_{ij}(\mathbf{z},\mathbf{x})\,u_j(\mathbf{x})\,\mathrm{d}\Gamma(\mathbf{x}) = \int_\Gamma u^\star_{ij}(\mathbf{z},\mathbf{x})\,t_j(\mathbf{x})\,\mathrm{d}\Gamma(\mathbf{x}) + \int_\Omega u^\star_{ij}(\mathbf{z},\mathbf{x})\,f_j(\mathbf{x})\,\mathrm{d}\Omega(\mathbf{x}), \tag{1}$$

where $u_j(\mathbf{x})$ and $t_j(\mathbf{x})$ are the j-th ($j = 1 \ldots N_d$) components of the displacement (flow potential) and traction (flux) boundary fields of dimensionality N_d ($N_d = 1$ for Potential Problems and N_d=2 or 3 for two- or three-dimensional Elastostatics); $f_j(\mathbf{x})$ denotes a distribution of known body forces (sources). In this equation $u^\star_{ij}(\mathbf{z},\mathbf{x})$ and $t^\star_{ij}(\mathbf{z},\mathbf{x})$ represent the Kelvin's fundamental solution, being $\mathbf{z}$ the load (or source) point and $\mathbf{x}$ the field point, both lying on the boundary Γ of the domain Ω. As it is well known, the kernel functions have the following order of singularities when $\mathbf{x}$ approaches $\mathbf{z}$: $u^\star_{ij}(\mathbf{z},\mathbf{x}) = \mathrm{O}(\log r)$ and $t^\star_{ij}(\mathbf{z},\mathbf{x}) = \mathrm{O}(1/r)$, with $r = |\mathbf{z}-\mathbf{x}|$. Then, the first integral in (1) must be understood in the Cauchy Principal Value (CPV) sense. If eq. (1) is considered on boundary points, $\mathbf{z} \in \Gamma$, the free-term $c_{ij}(\mathbf{z})$ depends on the geometry of the boundary at $\mathbf{z}$ and is equal to $1/2\,\delta_{ij}$ (δ_{ij} is the Kronecker symbol) when the geometry is sufficiently "smooth", otherwise, closed form expressions for the free-term are presented by Hartmann[15].

Introducing the prescribed boundary conditions

$$u_j = \bar{u}_j, \text{(on } \Gamma_u\text{)}, \quad \text{and} \quad t_j = \bar{t}_j, \text{(on } \Gamma_t\text{)}, \tag{2}$$

eq. (1) leads to

$$\begin{cases} c_{ij}\,u_j + \fint_{\Gamma_t} t^\star_{ij}\,u_j\,\mathrm{d}\Gamma - \int_{\Gamma_u} u^\star_{ij}\,t_j\,\mathrm{d}\Gamma - b_i = 0, \text{(on } \Gamma_t\text{)}; \\ \int_{\Gamma_t} t^\star_{ij}\,u_j\,\mathrm{d}\Gamma - \int_{\Gamma_u} u^\star_{ij}\,t_j\,\mathrm{d}\Gamma - b_i = 0, \text{(on } \Gamma_u\text{)}, \end{cases} \tag{3}$$

where

$$\begin{cases} b_i = -\int_{\Gamma_u} t^\star_{ij}\,\bar{u}_j\,\mathrm{d}\Gamma + \int_{\Gamma_t} u^\star_{ij}\,\bar{t}_j\,\mathrm{d}\Gamma + \int_\Omega u^\star_{ij}\,f_j\,\mathrm{d}\Omega, \text{(on } \Gamma_t\text{)}; \\ b_i = -c_{ij}\,\bar{u}_j - \fint_{\Gamma_u} t^\star_{ij}\,\bar{u}_j\,\mathrm{d}\Gamma + \int_{\Gamma_t} u^\star_{ij}\,\bar{t}_j\,\mathrm{d}\Gamma + \int_\Omega u^\star_{ij}\,f_j\,\mathrm{d}\Omega. \text{(on } \Gamma_u\text{)}. \end{cases} \tag{4}$$

In these equations and hereinafter, subscripts ($\mathbf{z}$), ($\mathbf{x}$) and ($\mathbf{z}\ \mathbf{x}$) are dropped for the sake of brevity.

For the general case of boundary mixed problems, eqs (1) and (3) involve boundary unknowns which are dimensionally different. Hence, to obtain variables in the same range, a suitable scaling is convenient. For instance, in elastostatic problems shear modulus G can be used to scale tractions and forces, by

$$t_j = t_j/G, \quad \text{and} \quad f_j = f_j/G, \tag{5}$$

becoming the kernel displacements

$$u^\star_{ij} = G\,u^\star_{ij}, \tag{6}$$

where the equal sign is used in the sense of the Fortran statement of equivalence to avoid introducing additional variables.

To obtain a solution by a boundary element process, the physical boundary unknowns must be approximated through a suitable set of interpolation functions φ^k, by

$$u_i \simeq \hat{u}_i = \sum_k \varphi^k\,\hat{u}^k_i, \text{(on } \Gamma_t\text{)}, \quad \text{and} \quad t_i \simeq \hat{t}_i = \sum_k \varphi^k\,\hat{t}^k_i, \text{(on } \Gamma_u\text{)}. \tag{7}$$

The difference between the exact and approximated BEM solutions can be defined through the error functions

$$u^e_i = u_i - \hat{u}_i, \text{(on } \Gamma_t\text{)}, \quad \text{and} \quad t^e_i = t_i - \hat{t}_i, \text{(on } \Gamma_u\text{)}, \tag{8}$$

which substituted in the boundary integral statement (3), leads to the "residual displacement" function[9,10]

$$\begin{cases} r_i = c_{ij}\,\hat{u}_j + \fint_{\Gamma_t} t^\star_{ij}\,\hat{u}_j\,\mathrm{d}\Gamma - \int_{\Gamma_u} u^\star_{ij}\,\hat{t}_j\,\mathrm{d}\Gamma - b_i, \text{(on } \Gamma_t\text{)}; \\ r_i = \int_{\Gamma_t} t^\star_{ij}\,\hat{u}_j\,\mathrm{d}\Gamma - \int_{\Gamma_u} u^\star_{ij}\,\hat{t}_j\,\mathrm{d}\Gamma - b_i, \text{(on } \Gamma_u\text{)}. \end{cases} \tag{9}$$

which, written in order of the error functions (8) has the form

$$\begin{cases} r_i = -c_{ij}\,u^e_j - \fint_{\Gamma_t} t^\star_{ij}\,u^e_j\,\mathrm{d}\Gamma + \int_{\Gamma_u} u^\star_{ij}\,t^e_j\,\mathrm{d}\Gamma, \text{(on } \Gamma_t\text{)}; \\ r_i = -\int_{\Gamma_t} t^\star_{ij}\,u^e_j\,\mathrm{d}\Gamma + \int_{\Gamma_u} u^\star_{ij}\,t^e_j\,\mathrm{d}\Gamma, \text{(on } \Gamma_u\text{)}. \end{cases} \tag{10}$$

SYSTEM OF EQUATIONS. COLLOCATION AND GALERKIN APPROACHES

To obtain each one of the degrees of freedom (d.o.f.) defined in interpolation (7), either the collocation or the Galerkin method can be used.

Following a collocation procedure, the residue r_i must be set to zero in as many nodes $\mathbf{z}^k$ of the boundary as the number of d.o.f. used to approximate the boundary physical unknowns. Hence, when indices i and k vary from 1 to N_i and from 1 to N_i (the total number of d.o.f. used to approximate i component of physical unknowns in all the boundary), respectively, the statement

$$r_i = 0 \tag{11}$$

leads to the standard system of equations

$$\mathbf{A}\mathbf{x} = \mathbf{b} \tag{12}$$

obtained in the BEM.

If a Galerkin approach is used, the residue is "weighted' through the same interpolation functions used to approximate the unknowns A similar system of equations is obtained when the same variation of the indices i and k is considered in the statement

$$\int_{\Gamma} \varphi^k \, r_i \, \mathrm{d}\Gamma = 0. \tag{13}$$

Recentely, Parreira[16] and Parreira and Guiggiani[17] have presented a numerical integration scheme for the evaluation of double integrals arising in this approach.

HIERARCHIC INTERPOLATION

Hierarchic formulations had been successfully used in adaptive processes since 1975[18]. The main advantage of using such kinds of approximations is that when new d.o.f. are added to refine a certain mesh, the matrix $\mathbf{A}$ and the vector b used in the previous system of equations remain unchanged.

For plane problems and the p-version, the most common families of interpolation functions used are the Legendre polynomials[19] and the Peano's family[18].

Although the first one leads to matrices with lower condition numbers[9,10], the use of the latter seems to be more convenient in the BEM. That is because: i) as shown by Parreira[9] and Mullen and Rencis[21], the use of iterative methods for solving linear systems, the convergence of which is strongly dependent on the condition number of $\mathbf{A}$, are not recommendable in plane problems; *ii*) if a continuous model is to be used, interelement continuity is easier to be implemented using Peano's family; *iii*) practical experience tells that a BEM solution is strongly dependent on the accuracy of the numerical integration process. Due to their oscillatory character, Legendre polynomials (as Lagrange functions) lead to larger numerical integration errors.

ERROR INDICATORS AND ESTIMATORS

In boundary elements error indicators and estimators become defined by different researchers using the residue in two alternative ways.

Like Rank[3,13] Umetani[7] or Parreira[9,10], the residue can be used directly to quantify the error of the solution either in a H^0 norm

$$\| e \|_0^2 = \int_\Gamma \sum_i r_i^2 \, \mathrm{d}\Gamma, \tag{14}$$

or in a H^1 norm

$$\| e \|_1^2 = \int_\Gamma \sum_i \left[r_i^2 + \left(\frac{\partial r_i}{\partial s} \right)^2 \right] \mathrm{d}\Gamma. \tag{15}$$

The use of norms like the one defined in (15) do not present any kind of evident advantages in the BEM over the one defined in (14). However, if one wants to use them, it must be done with special care, because the quantities r_i and $\partial r_i / \partial s$ are dimensionally different. Hence, to add them, some kind of scaling is needed. Moreover, computation of ∂r_i ∂s involves the evaluation of integrals of high singular kernels which must be interpreted in the finite-part sense[20].

Using definition (14), an error estimator η can be obtained from error indicators λ_i^k of all the elements by

$$\eta^2 = \| e \|_0^2 = \int_\Gamma \sum_i r_i^2 \, \mathrm{d}\Gamma = \sum_{i,e} \int_{\Gamma_e} r_i^2 \, \mathrm{d}\Gamma = \sum_{i,e} (\lambda_i^e)^2 , \tag{16}$$

being

$$(\lambda_i^e)^2 = \int_{\Gamma_e} r_i^2 \, \mathrm{d}\Gamma. \tag{17}$$

Alternatively, Alarcon and his co-workers[4,5,6,8,11,12], following the finite element developments, use a so called "energy norm" defined as[8]

$$\| e \|_E^2 = - \int_\Gamma \sum_i e_i r_i \, \mathrm{d}\Gamma, \tag{18}$$

where e_i is the error function defined in (8). Note that both e_i and r_i have the dimension of displacements, so, the integrand function of eq. (18) needs extra scaling to represent an energy functional.

Error indicators can now be obtained considering that in a certain level of the adaptive process, the error in a given element Γ_e is approximated by the hierarchical interpolation function φ^{n+1} of next degree[8,11,12].

$$e_i \simeq \varphi^{n+1} e_i^{n+1}. \tag{19}$$

Substituting eq. (19) in definition (18), one has

$$\| e \|_E^2 = \int_{\Gamma_e} \varphi^{n+1} r_i \, d\Gamma \, e_i^{n+1}. \tag{20}$$

Substituting now eq. (19) in eq. (10), and afterwards in definition (18), one has

$$\begin{cases} \| e \|_E^2 = \left[\int_{\Gamma_e} \varphi^{n+1} c_{ij} \varphi^{n+1} \, d\Gamma + \iint_{\Gamma_e} \varphi^{n+1} t_{ij}^{\star} \varphi^{n+1} \, d\Gamma \, d\Gamma \right] (e_i^{n+1})^2, \\ \text{(on } \Gamma_t\text{);} \\ \| e \|_E^2 = \left[- \iint_{\Gamma} \varphi^{n+1} u_{ij}^{\star} \varphi^{n+1} \, d\Gamma \, d\Gamma \right] (e_i^{n+1})^2, \text{(on } \Gamma_u\text{).} \end{cases} \tag{21}$$

Comparing eqs. (20) and (21) and afterwards substituting the result obtained for e_i^{n+1} in eq. (20) or (21), one can obtain for the error indicators the result

$$\begin{cases} (\lambda_i^e)^2 = \dfrac{\left(\int_{\Gamma_e} \varphi^{n+1} r_i \, d\Gamma \right)^2}{\int_{\Gamma_e} \varphi^{n+1} c_{ij} \varphi^{n+1} \, d\Gamma + \iint_{\Gamma_e} \varphi^{n+1} t_{ij}^{\star} \varphi^{n+1} \, d\Gamma \, d\Gamma}, \text{(on } \Gamma_t\text{);} \\ (\lambda_i^e)^2 = \dfrac{\left(\int_{\Gamma_e} \varphi^{n+1} r_i \, d\Gamma \right)^2}{- \iint_{\Gamma_e} \varphi^{n+1} u_{ij}^{\star} \varphi^{n+1} \, d\Gamma \, d\Gamma}, \text{(on } \Gamma_u\text{).} \end{cases} \tag{22}$$

If a Galerkin approach is used, at least in higher order degrees of interpolation, it can easily be proved[9] that

$$\int_{\Gamma_e} \varphi^{n+1} r_i \, d\Gamma = \sum_l A_{n+1,l} \, x_l - b_{n+1}. \tag{23}$$

Considering that

$$\begin{cases} \int_{\Gamma_e} \varphi^{n+1} c_{ij} \varphi^{n+1} \, d\Gamma + \iint_{\Gamma_e} \varphi^{n+1} t_{ij}^{\star} \varphi^{n+1} \, d\Gamma \, d\Gamma = A_{n+1,n+1}, \text{(on } \Gamma_t\text{);} \\ - \iint_{\Gamma_e} \varphi^{n+1} u_{ij}^{\star} \varphi^{n+1} \, d\Gamma \, d\Gamma = A_{n+1,n+1}, \text{(on } \Gamma_u\text{),} \end{cases} \tag{24}$$

the classical result obtained by Peano[18] for the finite element method.

$$(\lambda_i^e)^2 = \frac{\left(\sum_l A_{n+1,l}\, x_l - b_{n+1}\right)^2}{A_{n+1,n+1}}, \tag{25}$$

can be recovered from eq. (22).

For error estimators, following Gago[19], one can use

$$\eta^2 = \sum_{i,e} (\lambda_i^{\star e})^2, \tag{26}$$

where

$$(\lambda_i^{\star e})^2 = \frac{\int_{\Gamma_e} \left(\varphi^{n+1}\right)^2 \, d\Gamma}{A_{n+1,n+1}} \int_{\Gamma_e} r_i^2 \, d\Gamma \geq (\lambda_i^e)^2. \tag{27}$$

RESIDUAL FUNCTION

Using either eqs. (16) and (17) or eqs. (22) and (27) for the evaluation of error indicators and estimators one must integrate the residue along the boundary. This function can be directly calculated in all points throught eq. (9), as soon as the boundary element solution is known.

The main problem in computing directly the integrals of the residue consists in evaluating it in points too close to the elements ends. To calculate the residue very sophisticated numerical integration schemes to obtain CPV integrals[23,24] and quasi-singular integrals[22] have been used. Nevertheless, as it can be seen in fig. 4, the residue presents strange oscillations in the ends, from which it can be concluded that the obtained values are "wrong".

When a discontinuous model is used, the residue presents singularities of order O(log r) in the points where the displacement BEM solution is discontinuous[9]. In itself, this does not represent any inconvenience because this kind of singularity can be easily removed using a convenient co-ordinate transformation[22]. The problem when using this kind of self-adaptive transformations is that it moves the Gauss points towards the singularity, e.g. to the elements ends. So, the probability that a Gauss point falls in the "dangerous zone" becomes greater, which when it happens will adulterate completely the obtained result. Moreover, as the above mentioned integration schemes give very accurate results when the point considered is an element end, the use of continuous models instead of discontinuous ones is more convenient, because it becomes possible to use Lobatto's formulas[27], which have the advantage over the traditional Gauss-Legendre ones of using interior points further away from the ends.

COLLOCATION, GALERKIN OR MIXED COLLOCATION-GALERKIN APPROACHES IN ADAPTIVE BEM

The use of collocation or Galerkin methods in Adaptive BEM were discussed by Umetani[7] and Parreira[9], who concluded that Galerkin approach is more appropriate than the collocation one. The main reason for that is the choice of the position for the nodes inside the elements. They can not be in the optimal position[25,26] because it is different when linear, parabolic, cubic, etc. interpolations are considered. However, one can choose, for instance, the elements ends for the nodes associated with the linear interpolation, which is an obvious choice. Then, associated with parabolic function the 1/2 point can be chosen, which is obvious too. But which node must be associated to the cubic function? It can be chosen the 1/4 point but, this choice will introduce a big distortion in the residual function, which can even make the solution itself conditional.

Now, it is possible to add some more advantages to the ones just mentioned. First, if the Galerkin method is used, it will be possible to use definition (25) for error indicators which has the advantage over the ones defined in (22) that some of the calculation then made, as for $A_{n+1,l}$, $A_{n+1,n+1}$ and b_{n+1} will be used in the system of equations of the preceding step. Besides, if collocation is used, to evaluate the integral of the residue over all the boundary more Gauss (or Lobatto) points will be needed.

However, continuous models are much easier to be implemented using the collocation model. Hence, a mixed approach, which gathers the advantages of both methods can be proposed. This approach uses the collocation method with nodes at elements ends in the linear initial mesh and Galerkin approach for higher order interpolations. Besides this, as the residue is zero in all the elements ends, it must be computed only at interior points when Lobatto's formulas are used.

INDIRECT EVALUATION OF THE RESIDUE

A direct numerical integration of the residue as the one discussed in the previous sections presents some disadvantages: *i*) it involves a very high computational effort, *ii*) it requires very sophisticated integration schemes, *iii*) it is always possible that an integration point falls in the "dangerous zone". Alternatively to its direct evaluation, the residual function integrals can be obtained in an indirect way.

Let it be considered that its component i in the element Γ_e is approximated through the same set of interpolation functions used for the physical unknowns. Then,

$$r_i = \sum_{l=0}^{m} \varphi^l \, r_i^l, \tag{28}$$

where m is the degree until which one wants to consider the mentioned approximation, which must be greater than the degree n of the current approximation used for the physical unknowns.

The $m+1$ d.o.f. of eq. (28) can be obtained using a Galerkin type procedure. Hence, statement

$$\sum_{l=0}^{m} \int_{\Gamma_e} \varphi^k \varphi^l \, \mathrm{d}\Gamma \; r_i^k = \int_{\Gamma_e} \varphi^k \, r_i \, \mathrm{d}\Gamma \tag{29}$$

leads to the symmetric linear system of equations

$$\mathbf{C}\,\mathbf{r} = \mathbf{d} \tag{30}$$

when the upper index k goes from 0 to m. Note that from eq. (13) all the d_k, with $k = 2 \ldots n$ (or with $k = 0 \ldots n$, in a discontinuous model), are zero. Moreover, from eq. (23) it is clear that a lot of the computations used to obtain d_k (with $k > n$) can be used to form the system of equations of the next steps of an adaptive process.

The eq. (30) can be re-written in the form

$$\begin{bmatrix} \mathbf{C}_{11} & \mathbf{C}_{12} & \mathbf{C}_{13} \\ \mathbf{C}_{21} & \mathbf{C}_{22} & \mathbf{C}_{23} \\ \mathbf{C}_{31} & \mathbf{C}_{32} & \mathbf{C}_{33} \end{bmatrix} \begin{bmatrix} \mathbf{r}_1 \\ \mathbf{r}_2 \\ \mathbf{r}_3 \end{bmatrix} = \begin{bmatrix} \mathbf{d}_1 \\ \mathbf{d}_2 \\ \mathbf{d}_3 \end{bmatrix}, \tag{31}$$

where

$$\mathbf{r}_1 = \begin{bmatrix} r_0 \\ r_1 \end{bmatrix}, \; \mathbf{r}_2 = \begin{bmatrix} r_2 \\ \vdots \\ r_n \end{bmatrix} \; \text{and } \mathbf{r}_3 = \begin{bmatrix} r_{n+1} \\ \vdots \\ r_m \end{bmatrix}, \tag{32}$$

which, after some trivial arithmetic manipulations not included here for brevity, leads to the result

$$\begin{cases} \mathbf{r}_1 = \mathbf{D}_{11}^{-1} \left(\mathbf{d}_1 - \mathbf{D}_{13} \mathbf{r}_3 \right) \\ \mathbf{r}_2 = -\mathbf{C}_{22}^{-1} \left(\mathbf{C}_{21} \mathbf{r}_1 + \mathbf{C}_{23} \mathbf{r}_3 \right) \\ \mathbf{r}_3 = \mathbf{D}_{33}^{\star\,-1} \left(\mathbf{d}_3 - \mathbf{D}_{31} \mathbf{D}_{11}^{-1} \mathbf{d}_1 \right) \end{cases}, \tag{33}$$

where

$$\mathbf{D}_{\alpha\beta} = \mathbf{C}_{\alpha\beta} - \mathbf{C}_{\alpha 2} \mathbf{C}_{22}^{-1} \mathbf{C}_{2\beta} \quad (\alpha, \beta = 1 \text{ or } 3) \tag{34}$$

and

$$\mathbf{D}_{33}^{\star} = \mathbf{D}_{33} - \mathbf{D}_{31} \mathbf{D}_{11}^{-1} \mathbf{D}_{13}. \tag{35}$$

Considering eqs. (30) and (33) and that $\mathbf{d}_2 = \mathbf{0}$, eq. (17) written in a matrix notation leads to final result for the error indicators

$$\begin{aligned}(\lambda_i^e)^2 &= \mathbf{r}^T\,\mathbf{C}\,\mathbf{r} = \mathbf{r}^T\,\mathbf{d} = \mathbf{d}^T\,\mathbf{r} = \mathbf{d}_1^T\,\mathbf{r}_1 + \mathbf{d}_2^T\,\mathbf{r}_2 + \mathbf{d}_3^T\,\mathbf{r}_3 = \mathbf{d}_1^T\,\mathbf{r}_1 + \mathbf{d}_3^T\,\mathbf{r}_3 = \\ &= \bar{\mathbf{d}}^T\,\bar{\mathbf{D}}\,\bar{\mathbf{d}},\end{aligned} \tag{36}$$

where

$$\bar{\mathbf{d}} = \begin{bmatrix} \mathbf{d}_1 \\ \mathbf{d}_3 \end{bmatrix} \tag{37}$$

and

$$\bar{\mathbf{D}} = \begin{bmatrix} \mathbf{D}_{11}^{-1} - \mathbf{D}_{11}^{-1}\,\mathbf{D}_{13}\,\mathbf{D}_{33}^{\star\,-1}\,\mathbf{D}_{31}\,\mathbf{D}_{11}^{-1} & -\mathbf{D}_{11}^{-1}\,\mathbf{D}_{13}\,\mathbf{D}_{33}^{\star\,-1} \\ -\mathbf{D}_{33}^{\star\,-1}\,\mathbf{D}_{31}\,\mathbf{D}_{11}^{-1} & \mathbf{D}_{33}^{\star\,-1} \end{bmatrix} \tag{38}$$

Note that $\bar{\mathbf{D}}$ is a symmetric matrix of order $m - n + 2$ (or $m - n$, in a discontinuous model) and it is calculated once for all if the simplification

$$C_{kl} = \int_{\Gamma_e} \varphi^k\,\varphi^l\;\mathrm{d}\Gamma = \int_{-1}^{+1} \varphi^k(\xi)\,\varphi^l(\xi)\,\mathrm{J}(\xi)\;\mathrm{d}\xi \simeq L_e \int_{-1}^{+1} \varphi^k(\xi)\,\varphi^l(\xi)\;\mathrm{d}\xi, \tag{39}$$

being

$$L_e = \int_{-1}^{+1} \mathrm{J}(\xi)\;\mathrm{d}\xi, \tag{40}$$

is used.

NUMERICAL EXAMPLES

To test the indirect method to evaluate the residue over the boundary, through which error indicators and estimators can be obtained, three classic plane stress problems were analysed: i) the square plate with a parabolic compression load (fig. 1); ii) the square cantilever transversely loaded (fig. 2); iii) the square plate uniformly tractioned with a crack (fig. 3). In all the three cases, $E = 1$ and $\nu = 0$ were assumed. Each example was analysed with two meshes, the 4, 4, and 5 elements mesh represented in each one of the mentioned figs. and the 8, 8, 10 elements mesh obtained from the former dividing each element in two. Each mesh was analysed using linear, parabolic and cubic interpolations.

In the tables 1 to 3, the error estimators obtained using either the direct or the indirect methods are presented. When the indirect method was used, the errors were obtained approximating the residue with a order of interpolation m one, two and three degrees more than the degree n used in the unknowns interpolation, e.g. $m - n = 1$, 2 or 3. That is, if, for instance, parabolic interpolation of the unknowns was considered than the errors were computed approximating the residue with cubic, quartic and quintic interpolations, alternatively.

The results contained in these tables strongly encourage the use of the indirect method to compute error indicators and estimators in adaptive BEM codes. Furthermore, the values obtained for them when $m - n = 1$ are accurate enough, so, this result is proposed to be used in practical applications.

CONCLUSIONS

The use of the Legendre polynomials or the Peano's family in p-hierarchic BEM is discussed and arguments are presented which point to the use of the Peano's family

The shape of the residual function along the boundary obtained by the collocation or the Galerkin approaches, using either a continuous or a discontinuous model, is analysed. A mixed collocation-Galerkin method, which uses the collocation method in the linear interpolation and the Galerkin method for higher order interpolations, is proposed.

An indirect method for the evaluation of residual integrals, needed to compute the error indicators and estimators is presented. This method leads to the quadratic form (36), where the symmetric matrix $\bar{\mathbf{D}}$ is a-priori known and is presented in appendix when $m - n$ is considered equal to 1.

REFERENCES

1. Zienkiewicz, O.C., Gago, J.P. de S.R. and Kelly, D.W. (1983), The hierarchical concept in finite element analysis, Comp. and Struct., **16**, No. 1-4.

2. Alarcon, E., Avia, L. and Reverter, A. (1984), On the possibility of adaptive boundary elements, Proc. of the Int. Conf. on Accuracy Estimates and Adaptative Refinements in Finite Element Comp. (ARFEC), Lisboa.

3. Rank, E. (1984), A posteriori error estimates and adaptative refinement for some boundary integral element methods, Proc. of the Int. Conf. on Accuracy Estimates and Adaptative Refinements in Finite Element Comp. (ARFEC), Lisboa.

4. Alarcon, E., Reverter, A. and Molina (1985), J., Hierarchical boundary elements, Comput. and Struct. **20**, No. 1-3, 151–156.

5. Alarcon, E. and Reverter, A. (1985), p-Adaptive boundary elements, Int. Jou. Num. Meth. in Engng., **23**(1-3), 151–156.

6. Reverter, A., Gonzalez, A. and Alarcon, E. (1985), Indicators and estimators in P-adaptive boundary elements, Proc. 7th Int. Conf. on

Boundary Elements. (C.A. Brebbia and G. Maier Eds.). Springer-Verlag. Berlin. 12.41–12.54.

7. Umetani. S. (1986). Reliability and Efficiency of the Boundary Element Method in Elastostatics. Ph.D. Thesis. Univ. of Southampton.

8. Cerrolaza. M. and Alarcon. E. (1987). p-Adaptive boundary elements for three dimensional potential problems. in Comm. in Appl. Num. Meth. Vol. 3. 335-244.

9. Parreira P. (1987). Análise do Erro no Método dos Elementos de Fronteira em Elasticidade. Tese de Doutoramento. Univ. Técnica de Lisboa. Portugal. (in portughese).

10. Parreira. P. (1987). Self-adaptive p-hierarchical boundary elements in elastostatics. in Boundary Element IX (C.A. Brebbia. W.L. Wendland. G. Kuhn. Ed.). Vol. 1. 351–373. Springer-Verlag. Stuttgart.

11. Gomez-Lera. S.. Cerrolaza. M. and Alarcon. E. (1987). Adaptive refinements in BEM. in Boundary Element IX (C.A. Brebbia. W.L. Wendland. G. Kuhn. Ed.). Vol. 1. 337–349. Springer-Verlag. Stuttgart.

12. Cerrolaza. M. (1988). Elementos de Contorno p-adaptables: Desarollo y Aplicaciones en Teoria del Potencial y Elastostática. Tesis Doctoral. Univ. de Madrid. (in spanish).

13. Rank. E. (1987). Adaptive boundary element methods. in Boundary Element IX (C.A. Brebbia. W.L. Wendland. G. Kuhn. Ed.). Vol. 1. 259–278. Springer-Verlag. Stuttgart.

14. Brebbia C.A.. Telles J.C.F. and Wrobel L.C. (1984). Boundary Element Techniques: Theory and Applications in Engineering. Springer-Verlag. Berlin · Heidelberg.

15. Hartmann. F. (1980). Computing the C-matrix on non-smooth boundary points. in New Developments in Boundary Element Methods (C.A. Brebbia Ed.). Butterworths. London. 367-379.

16. Parreira. P. (1988). A numerical integration scheme for the Galerkin approach in Boundary Elements. in Boundary Element X (C.A. Brebbia Ed.). Vol. 1. 297–311. Springer-Verlag. Southampton.

17. Parreira. P. and Guiggiani. M. (1989). On the implementation of the Galerkin approach in the boundary element method. presented for publication on Computers & Structures

18. Peano, A.G. (1975). Hierarchies of conforming finite elements, thesis presented in partial fulfillment of the requirements for the Ph. D. degree, Washington Univ.

19. Gago, J.P. de S.R. (1982). A posteriori Error Analysis and Adaptivity for the Finite Element Method, Ph.D. Thesis, Univ. of Swansea.

20. Ingber, M.S., Mitra, A.K. and Rudolphi, T.J. (1988), Grid redistribution based on measurable error indicators for the direct boundary element method, presented at BEM X, Southampton.

21. Mullen, R.L. and Rencis, J.J. (1987), Iterative methods for solving boundary element equations, Comp. and Struct., **25**, No. 5, 713–723.

22. Telles, J.C.F. (1987), A self-adaptive co-ordinate transformation for efficient numerical evaluation of general boundary element integrals, Int. Jou. Num. Meth. in Engng., **24**, 959–973

23. Guiggiani, M. and Casalini, P. (1987), Direct computation of Cauchy principal value integrals in advanced boundary elements, Int. Jou. Num. Meth. in Engng. **24**, 1711–1720.

24. Casalini, P. and Guiggiani, M. (1988), New developments in the boundary element method for plane and axisymmetric elasticity, in Advanced Boundary Element Methods, (T.A. Cruse, Ed.), Springer-Verlag, Berlin, 81–91.

25. Xu, J.M. and Brebbia, C.A. (1986), Optimum positions for the nodes in discontinuous boundary elements, em Proc. of the 8th Conf. on Boundary Elements, Tokyo.

26. Parreira, P. (1988), On the accuracy of continuous and discontinuous boundary elements, submitted for publication in Engineering Analysis.

27. Stroud, A.H. and Secrest, D. (1966), Gaussian Quadrature Formulas, Prentice-Hall, New York.

Square plate with a parabolic compression load

N^o elements and order of interpolation n	$[\|\| e \|\|]_{dir}$	$[\|\| e \|\|]_{ind} / [\|\| e \|\|]_{dir} \times 100$		
		$m-n=1$	$m-n=2$	$m-n=3$
4_1	2.942×10^{-3}	98.9	99.8	99.8
4_2	1.278×10^{-4}	99.0	99.7	99.7
4_3	1.765×10^{-6}	98.7	99.0	99.1
8_1	4.272×10^{-4}	99.6	99.9	99.9
8_2	1.721×10^{-6}	99.9	100.2	100.2
8_3	1.845×10^{-8}	95.0	95.6	99.5

Table 1: Error estimators obtained by the direct and by the indirect methods. Square plate with a parabolic compression load.

Square cantilever transversely loaded

N^o elements and order of interpolation n	$[\|\| e \|\|]_{dir}$	$[\|\| e \|\|]_{ind} / [\|\| e \|\|]_{dir} \times 100$		
		$m-n=1$	$m-n=2$	$m-n=3$
4_1	7.540×10^{-2}	98.0	99.6	99.7
4_2	7.253×10^{-3}	99.2	99.7	99.7
4_3	1.541×10^{-4}	75.8	88.4	88.6
8_1	1.612×10^{-2}	99.5	99.8	99.9
8_2	2.561×10^{-4}	98.4	99.5	99.6
8_3	2.402×10^{-5}	93.5	95.9	96.1

Table 2: Error estimators obtained by the direct and by the indirect methods. Square cantilever transversely loaded.

Square plate uniformly tractioned with a crack

N^o elements and order of interpolation n	$[\|\| e \|\|]_{dir}$	$[\|\| e \|\|]_{ind} / [\|\| e \|\|]_{dir} \times 100$		
		$m-n=1$	$m-n=2$	$m-n=3$
5_1	2.255×10^{-2}	98.3	99.6	99.7
5_2	3.418×10^{-3}	96.2	98.7	99.0
5_3	1.256×10^{-3}	91.9	95.9	96.2
10_1	5.270×10^{-3}	99.1	99.7	99.7
10_2	9.416×10^{-4}	96.8	98.8	99.0
10_3	3.411×10^{-4}	93.0	96.1	96.4

Table 3: Error estimators obtained by the direct and by the indirect methods. Square plate uniformly tractioned with a crack.

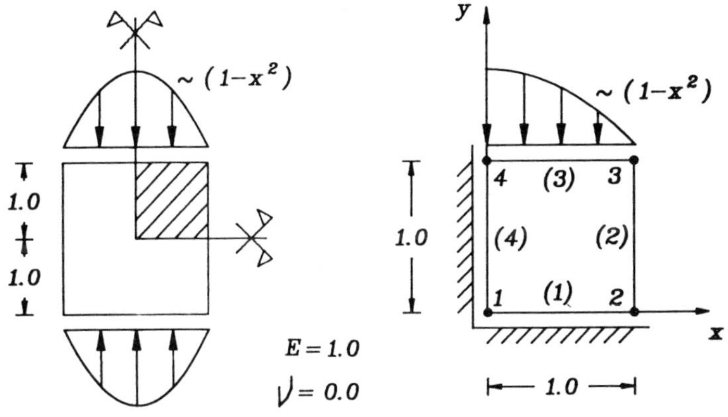

Figure 1: Square plate with a parabolic compression load.

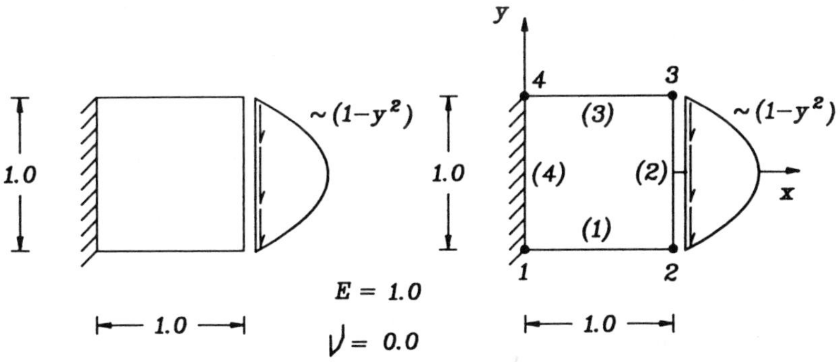

Figure 2: Square cantilever transversely loaded.

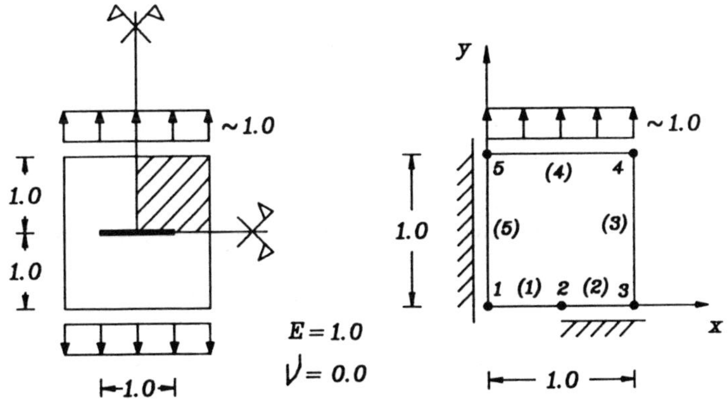

Figure 3: Square plate uniformly tractioned with a crack.

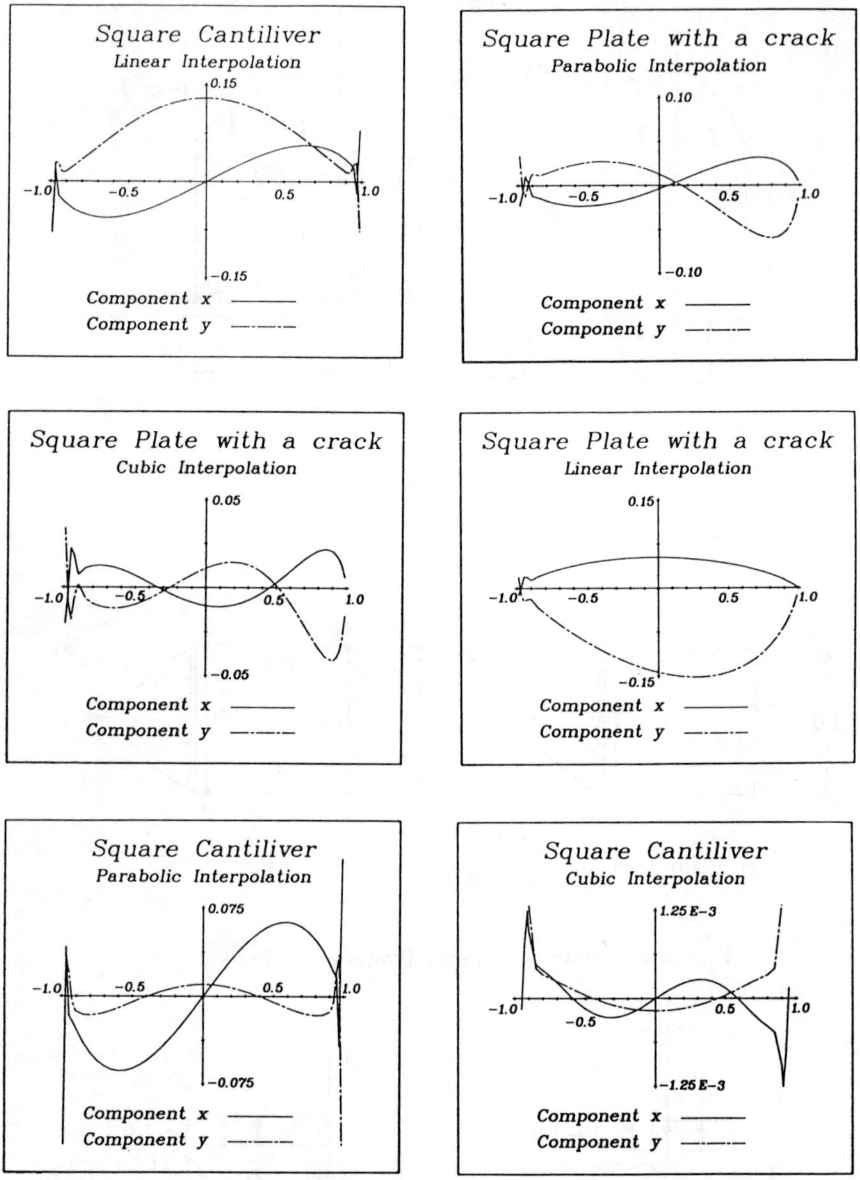

Figure 4: Residue on the element 2 of the square cantilever and the plate with a crack examples.

APPENDIX

The matrices $\bar{\mathbf{D}}_n$ used for the different degrees n of interpolation of the unknowns are the following:

$$\bar{\mathbf{D}}_1 = \frac{1}{L_e}\begin{bmatrix} 9 & 3 & -15/2 \\ & 9 & -15/2 \\ & & 45/4 \end{bmatrix}; \qquad \bar{\mathbf{D}}_2 = \frac{1}{L_e}\begin{bmatrix} 16 & -4 & 35/2 \\ & 16 & -35/2 \\ & & 175/4 \end{bmatrix};$$

$$\bar{\mathbf{D}}_3 = \frac{1}{L_e}\begin{bmatrix} 25 & 5 & -315/8 \\ & 25 & -315/8 \\ & & 11025/64 \end{bmatrix}; \qquad \bar{\mathbf{D}}_4 = \frac{1}{L_e}\begin{bmatrix} 36 & -6 & 693/8 \\ & 36 & -693/8 \\ & & 43659/64 \end{bmatrix};$$

$$\bar{\mathbf{D}}_5 = \frac{1}{L_e}\begin{bmatrix} 49 & 7 & -3003/16 \\ & 49 & -3003/16 \\ & & 693693/256 \end{bmatrix}; \quad \bar{\mathbf{D}}_6 = \frac{1}{L_e}\begin{bmatrix} 64 & -6 & 6435/16 \\ & 64 & -6435/16 \\ & & 2760615/256 \end{bmatrix}.$$

Over the past decade or so a great deal of work has been carried out to develop special techniques for determining accurate values of the stress intensity factor using the finite element method, such as the quarter-point elements [1,2], J-integral [3], strain energy release rate [4], virtual crack extension [6] and hybrid elements [7]. Many of these techniques have also been successfuly employed in boundary elements [8] as well as new techniques such as crack Green's function [9], displacement discontinuity technique [10] and subtraction of singularity technique [11]. Reviews of the techniques can be found in papers by Gallager [12] for finite elements and Smith [13] for boundary elements.

This paper concentrates on the application of most commonly used approaches in finite element method, namely quarter-point elements, virtual crack extension and J-integrals and assess their accuracy and efficiency against the boundary element technique of the subraction of the singularity [11].

3 Finite Elements

Applications of the finite element method to crack problems are well documented [12]. Here, methods of evaluating the stress intensity factors using quarter-point element, virtual crack extension technique and J-integrals are briefly described.

Crack-tip singularity modelling can be based on special crack-tip elements which directly model the $1/\sqrt{r}$ tip singularity, where r is the distance from the crack tip, as shown in figure 1. Following Henshell and Shaw [1] and Barsoum [2], this singularity

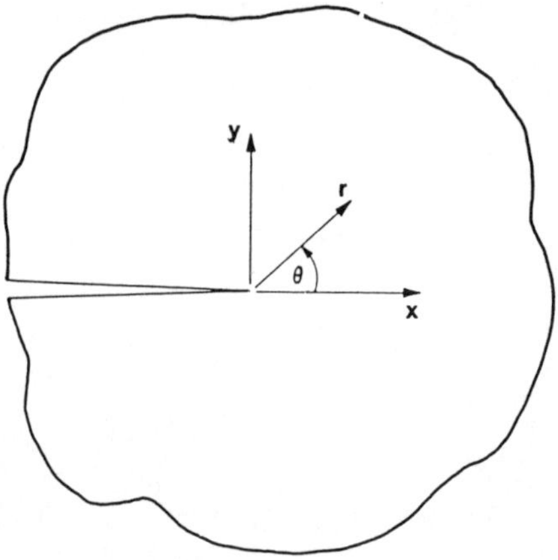

Figure 1: Crack geometry and polar coordinate system.

is most conveniently introduced into a quadratic isoparametric element by shifting the mid-side node to a quarter point position in the direction of the crack-tip, see figure 2. The stress intensity factor is calculated by equating the displacement field on the crack face approximated by the quarter-point element with the first term of the near-tip displacement field [14] to give, see figure 2:

$$K_I = \frac{\mu}{2(1-\nu)}\sqrt{\frac{2\pi}{l}}(4v_2 - v_3) \tag{1}$$

where ν is the Poisson ratio, μ is modulus of elasticity; l is the length of the quarter-point element, v_2 and v_3 are displacement values in y direction at nodes 2 and 3 respectively.

On the Accuracy of Boundary and Finite Element Techniques for Crack Problems in Fracture Mechanics

A. Portela and M.H. Aliabadi

Computational Mechanics Institute, Wessex Institute of Technology, Ashurst, Southampton SO4 2AA, UK

1 Abstract

Accuracy of boundary and finite element methods for the solution of linear elastic crack problems in Fracture Mechanics are examined. Special finite element techniques such as quarter-point elements, virtual crack extension and J-integral are compared against the boundary element singularity-subtraction technique.

Configurations studied for the comparison are, two symmetrical cracks emanating from a circular hole and a central crack, both in rectangular sheets subjected to tensile loading.

It is found that the boundary element singularity-subtraction technique yields more accurate values of the stress intensity factors with fewer degrees of freedom.

2 Introduction

The behaviour of cracks in structures is controlled by the stress intensity factor which is a function of the crack geometry and of the applied load. It is of importance to know the value of the stress intensity factor, since it is this factor which determines whether or not the crack will grow, and how fast it will grow. Analytical solutions for the stress intensity factor are limited to few idealised situations in which the domain is considered to be infinite and the loadings relatively simple. In practical situations however, the problems of fracture mechanics occur in flawed structures of finite size with complex loading. This has led to numerical approaches such as the finite element method and the boundary element method to be used for the solution of crack problems. The direct use of these methods involve computational difficuties in modelling the stress singularity at the crack-tip. Because of these difficulties they require either highly refined meshes around the tip and/or special techniques for representing the singularity and calculating stress intensity factors.

Over the past decade or so a great deal of work has been carried out to develop special techniques for determining accurate values of the stress intensity factor using the finite element method, such as the quarter-point elements [1,2], J-integral [3], strain energy release rate [4], virtual crack extension [6] and hybrid elements [7]. Many of these techniques have also been successfuly employed in boundary elements [8] as well as new techniques such as crack Green's function [9], displacement discontinuity technique [10] and subtraction of singularity technique [11]. Reviews of the techniques can be found in papers by Gallager [12] for finite elements and Smith [13] for boundary elements.

This paper concentrates on the application of most commonly used approaches in finite element method, namely quarter-point elements, virtual crack extension and J-integrals and assess their accuracy and efficiency against the boundary element technique of the subraction of the singularity [11].

3 Finite Elements

Applications of the finite element method to crack problems are well documented [12]. Here, methods of evaluating the stress intensity factors using quarter-point element, virtual crack extension technique and J-integrals are briefly described.

Crack-tip singularity modelling can be based on special crack-tip elements which directly model the $1/\sqrt{r}$ tip singularity, where r is the distance from the crack tip, as shown in figure 1. Following Henshell and Shaw [1] and Barsoum [2], this singularity

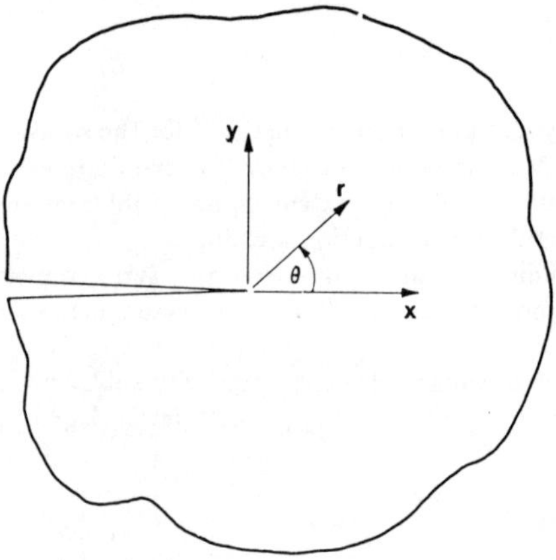

Figure 1: Crack geometry and polar coordinate system.

is most conveniently introduced into a quadratic isoparametric element by shifting the mid-side node to a quarter point position in the direction of the crack-tip, see figure 2. The stress intensity factor is calculated by equating the displacement field on the crack face approximated by the quarter-point element with the first term of the near-tip displacement field [14] to give, see figure 2:

$$K_I = \frac{\mu}{2(1-\nu)}\sqrt{\frac{2\pi}{l}}(4v_2 - v_3) \tag{1}$$

where ν is the Poisson ratio, μ is modulus of elasticity; l is the length of the quarter-point element, v_2 and v_3 are displacement values in y direction at nodes 2 and 3 respectively.

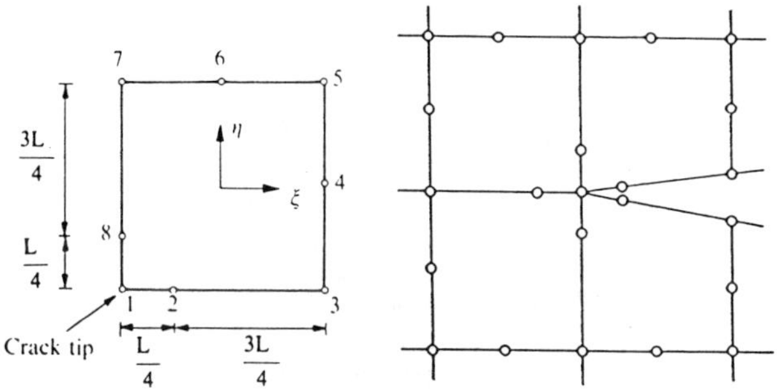

Figure 2: Crack-tip quarter-point finite element.

Alternatively, the finite element method can tackle crack problems using conservation laws that avoid the crack-tip singularity modelling. Under this alternative approach, the virtual crack extension technique, which is a variant of the energy release rate, relies upon a crack extension and was developed by Parks [5] and Hellen [6]. For a crack of length a, whose tip advances an amount δa, thereby causing a strain energy density release of amount δw, the strain energy release rate G is:

$$G = \frac{\delta w}{\delta a} \tag{2}$$

$$G = \frac{\eta + 1}{8\mu} K_I^2 \tag{3}$$

Following Hellen [6], the virtual crack-extension technique is performed in a finite element analysis, with an incremental crack length δa, in which the crack is extended as shown in figure 3, by advancing the nodal points within the crack-tip region only, in the direction of the crack axis. The amount of the crack-extension is recommended to lay between 0.001 and 0.1 of the crack-tip element length, [6]. The strain energy release rate is calculated from equation (2) and the stress intensity factor directly from equation (3).

The stress intensity factor is also related to path-independent integrals, known as J-integrals. For a contour as the one shown in figure 4, which encloses the crack-tip and has initial and end points on opposite surfaces of the crack, the Rice's [3] J-integral is defined as:

$$J = \int_{\Gamma} (W dy - t_i \frac{\partial u_i}{\partial x} d\Gamma) \tag{4}$$

where W denotes the strain energy density, u_i represents the displacements vector and t_i represents the tractions vector along the elementary arc $d\Gamma$ of the integration contour Γ. The J-integral is directly related to the stress intensity factor by the relation:

$$J = \frac{\eta + 1}{8\mu} K_I^2 \tag{5}$$

In a finite element analysis, the contour integration in equation (4) is carried out element-by-element, along a path through the numerical integration points, see figure 4. After the calculation of the J-integral, the evaluation of the stress intensity factor is straight forward made by equation (5).

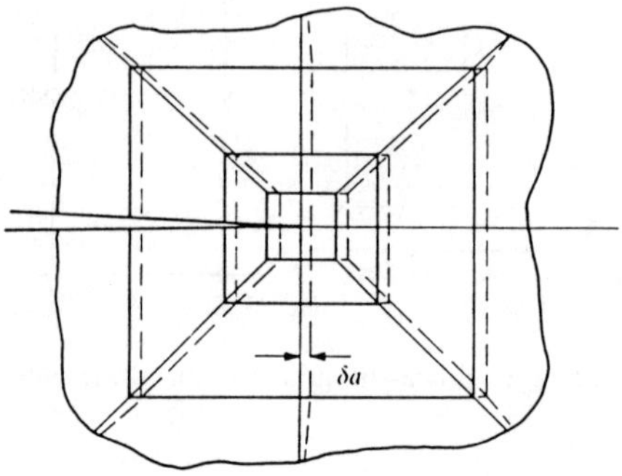

Figure 3: Finite element virtual crack extension.

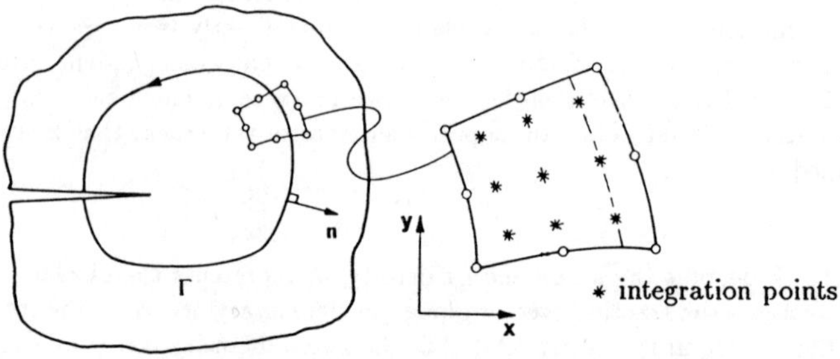

Figure 4: Contour path for finite element J-integral evaluation.

4 Boundary Elements

The boundary element method has been successfully applied to crack problems in Linear Elastic Fracture Mechanics (see references [13]). As in the finite element method, the direct application of the boundary element method requires highly refined meshes around the crack tip. Therefore, special approaches to deal with the crack-tip singularity must be used. Among all the boundary element formulations currently available for crack problems, only the singularity subtraction technique [11] will be considered in the present paper. Comparison between the singularity subtraction technique and other boundary element techniques are discussed by Smith and Aliabadi [15].

While most of other techniques incorporate the singular traction behavior into the boundary element analysis, the singularity subtraction technique removes it with a regularization procedure which is applied before the problem is solved. An important result of this procedure is the introduction of the stress intensity factor as one of the primary unknowns of the problem.

In a linear elastic region, containing a singularity, the displacement u and traction t fields can be separated into two components. For the traction field, these are a regular part, denoted by t^R and a singular part, denoted by t^S, while for the displacement field both components u^R and u^S are regular:

$$u = (u - u^S) + u^S = u^R + u^S \tag{6}$$

$$t = (t - t^S) + t^S = t^R + t^S \tag{7}$$

where u^S and t^S are obtained from eigenfunction expansions of the near-tip elastic field of a semi-infinite crack, [14]. Of course, the fields u^S and t^S automatically satisfy the governing equations and therefore the problem is now solved by the boundary element method in terms of the regular fields u^R and t^R, with the following boundary conditions:

$$\bar{u}_i^R = \bar{u}_i - \bar{u}_i^S \tag{8}$$

$$\bar{t}_i^R = \bar{t}_i - \bar{t}_i^S \tag{9}$$

These modified boundary conditions introduce additional unknowns, which are the coefficients of the referred series expansion. In order to obtain a unique solution for the final system of equations it is necessary to specify additional constraints, as many as the number of terms considered in the series expansion. These extra conditions are applied in such a way that the cancellation of the crack-tip singularity is ensured. This is done by directly constraining the traction field, *i.e.*, putting $t_i^R = 0$ at the crack-tip so that $t_i = t_i^S$. The stress intensity factor becomes one of the primary unknowns to be determined as the solution proceeds.

5 Configurations for Comparisons

To assess the performance of the quarter-point, virtual crack-extension and J-integral finite element approaches against the boundary element singularity-subtraction technique, the values obtained for the opening mode stress intensity factor are compared to alternative accurate results [16], for two different configurations of a rectangular plate of width W and height $H = 2W$ containing two symmetrical cracks of length $(a - R)$ emanating from a central circular hole of radius R, see figure 5. The plate is subjected to a uniaxial stress S applied on the ends. The first configuration studied was a central crack, $2R/W = 0.0$, with three different crack lengths, $2a/W = 0.3$, $2a/W = 0.4$ and $2a/W = 0.5$, respectively. The second configuration was a set of two cracks emanating from a central circular hole, $2R/W = 0.25$, with three different crack lengths, $2a/W = 0.3$, $2a/W = 0.4$ and $2a/W = 0.5$, respectively. Due to symmetry conditions, a quarter of the plate was only analysed for both configurations.

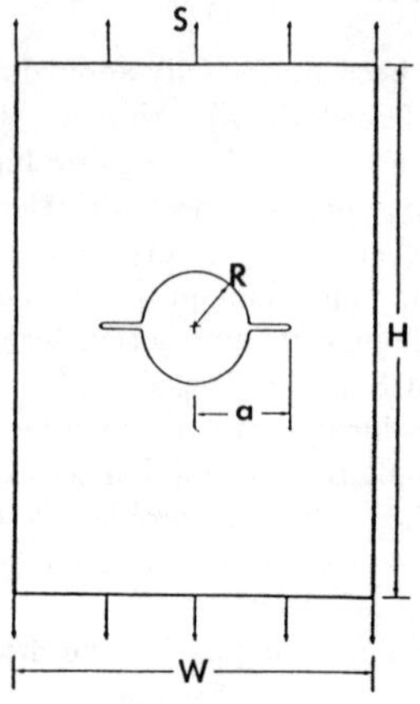

Figure 5: Plate Configurations for comparisons, $2R/W = 0.0$ and $2R/W = 0.25$.

6 Results for the Central Crack

In order to test the convergence of each method, three meshes were set up, for each crack length. The boundary element analysis was carried out on these three meshes with 14, 20 and 28 quadratic line elements, respectively, as shown in figure 6. For the finite element analysis three meshes with 19,60 and 68 quadratic quadrilateral elements were considered, respectively, as shown in figure 7. For both methods, finite elements and boundary elements, these meshes will be referred to as mesh-A, mesh-B and mesh-C, respectively. In order to keep the same mesh design for all crack lengths, mesh generation was used for both finite elements and boundary elements, in such a way that two corresponding meshes with different crack lengths would differ only in the near tip region. For the sake of simplicity the following notation will be used throughout the paper:

- BEM-SST: boundary elements, singularity-subtraction technique.
- FEM-JCI: finite elements, J-contour integral.
- FEM-VCE: finite elements, virtual crack-extension.
- FEM-QPE: quarter-point finite elements.

The results obtained for the central-crack problem are presented in tables 1 to 3 and plotted in figures 8 to 10. These tables present the values obtained for the stress intensity factors while the figures present the error obtained in the calculation of the stress intensity factors when compared with those values published by Newman [16], using a boundary coll ɔcation technique.

7 Cracks Emanating from a Circular Hole

Again, three meshes were set up for each crack length, to test the convergence. For this configuration, the boundary element analysis was carried out using 18, 26 and 36 quadratic elements, respectively, as shown in figure 11. For the finite element analysis three meshes with 19, 56 and 62 quadratic elements were considered, respectively, as shown in figure 12. For both methods, finite elements and boundary elements, these meshes will be referred to respectively as mesh-A, mesh-B and mesh-C. The results obtained for the

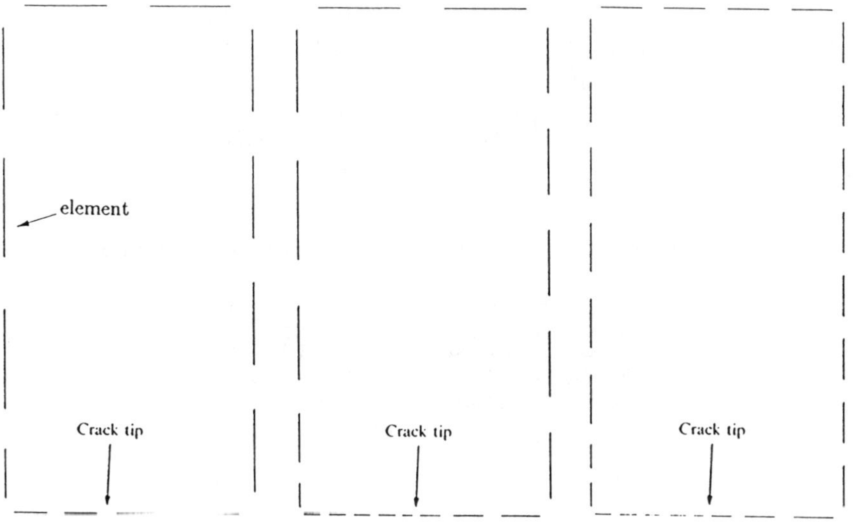

Figure 6: Boundary element mesh with quadratic elements; central crack; mesh-A, mesh-B and mesh-C, respectively.

Figure 7: Finite element mesh with quadratic elements; central crack; mesh-A, mesh-B and mesh-C, respectively.

$\frac{K_I}{S\sqrt{\pi a}}$	mesh-A	mesh-B	mesh-C
BEM-SST	1.075530	1.062106	1.059479
FEM-JCI	1.045793	1.065190	1.065757
FEM-VCE	1.046035	1.046030	1.057854
FEM-QPE	0.968787	0.986540	0.987031
$\frac{K_{ref16}}{S\sqrt{\pi a}}$	1.0583		

Table 1: Values of the stress intensity factor; central crack; $2a/W = 0.3$.

$\frac{K_I}{S\sqrt{\pi a}}$	mesh-A	mesh-B	mesh-C
BEM-SST	1.120976	1.112430	1.112290
FEM-JCI	1.056760	1.087410	1.087703
FEM-VCE	1.142614	1.089690	1.103870
FEM-QPE	1.187633	1.050040	1.050315
$\frac{K_{ref16}}{S\sqrt{\pi a}}$	1.1102		

Table 2: Values of the stress intensity factor; central crack; $2a/W = 0.4$.

$\frac{K_I}{S\sqrt{\pi a}}$	mesh-A	mesh-B	mesh-C
BEM-SST	1.192318	1.188438	1.188078
FEM-JCI	1.082850	1.137000	1.137304
FEM-VCE	1.139340	1.167370	1.164814
FEM-QPE	1.136875	1.133530	1.133840
$\frac{K_{ref16}}{S\sqrt{\pi a}}$	1.1876		

Table 3: Values of the stress intensity factor; central crack; $2a/W = 0.5$.

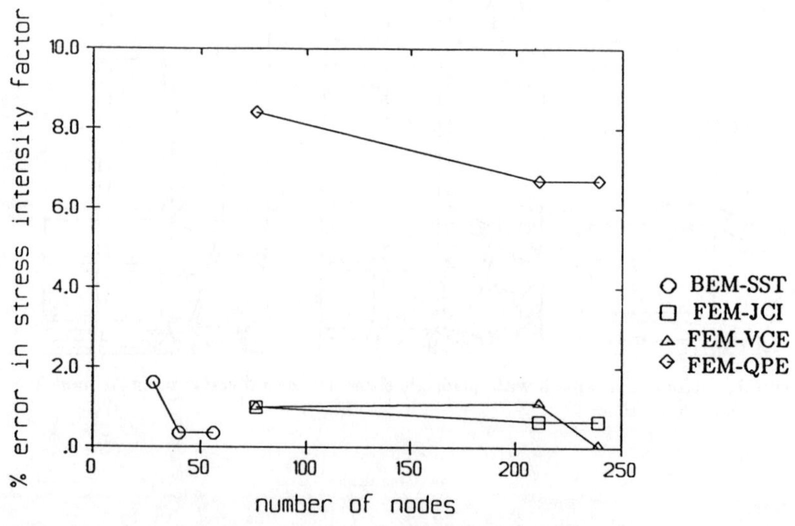

Figure 8: Error in the stress intensity factor calculation; central crack; $2a/W = 0.3$.

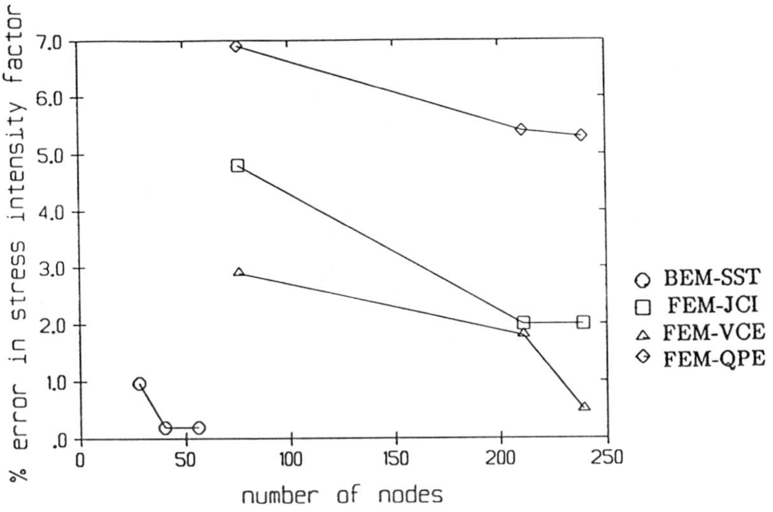

Figure 9: Error in the stress intensity factor calculation; central crack; $2a/W = 0.4$.

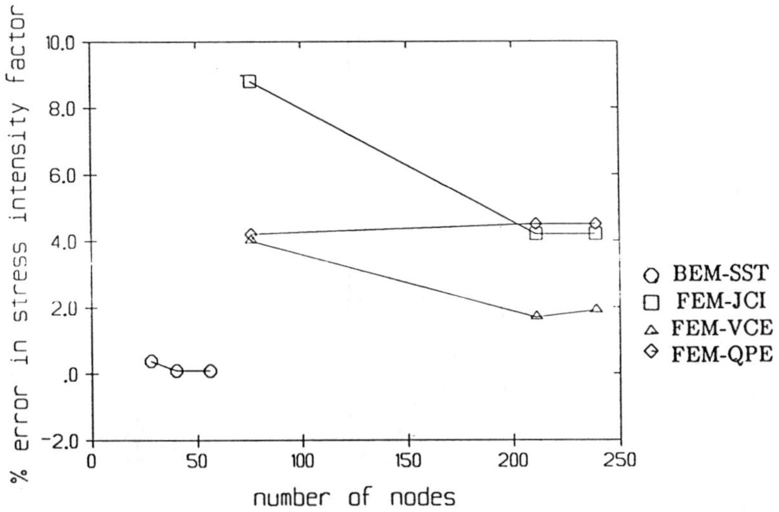

Figure 10: Error in the stress intensity factor calculation; central crack; $2a/W = 0.5$.

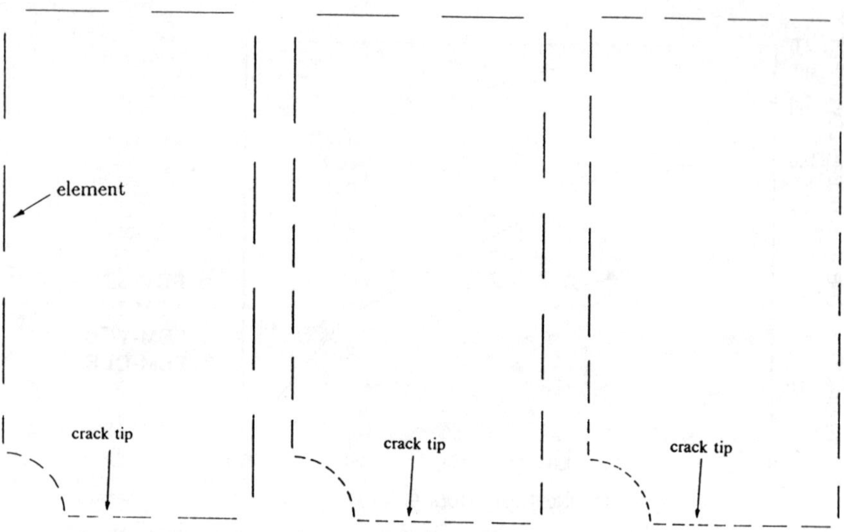

Figure 11: Boundary element mesh with quadratic elements; cracks emanating from a circular hole; $2R/W = 0.25$; mesh-A, mesh-B and mesh-C, respectively.

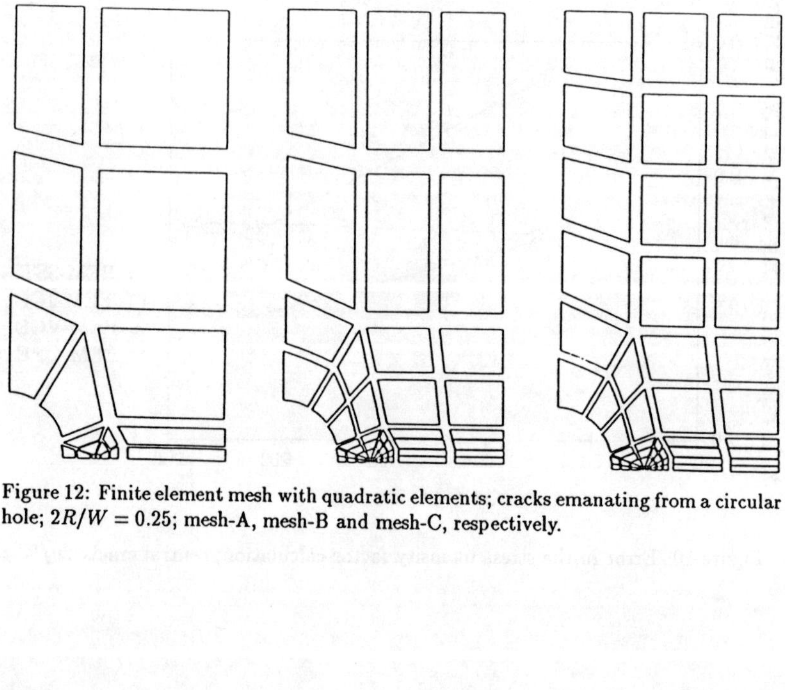

Figure 12: Finite element mesh with quadratic elements; cracks emanating from a circular hole; $2R/W = 0.25$; mesh-A, mesh-B and mesh-C, respectively.

problem of the cracks emanating from a circular hole are presented in tables 4 to 6 and plotted in figures 13 to 15. Once again, the tables present the values obtained for the stress intensity factors, while the figures present the error obtained in the calculation of the stress intensity factors when compared against those values obtained by Newman [16] with a boundary colocation method.

$\frac{K_I}{S\sqrt{\pi a}}$	mesh-A	mesh-B	mesh-C
BEM-SST	1.105874	1.078928	1.078319
FEM-JCI	1.055800	1.084200	1.085015
FEM-VCE	1.077200	1.061760	1.046035
FEM-QPE	0.992400	1.004600	1.005290
$\frac{K_{ref16}}{S\sqrt{\pi a}}$	1.0776		

Table 4: Values of the stress intensity factor; cracks emanating from a circular hole; $2R/W = 0.25$, $2a/W = 0.3$.

$\frac{K_I}{S\sqrt{\pi a}}$	mesh-A	mesh-B	mesh-C
BEM-SST	1.220081	1.216879	1.215987
FEM-JCI	1.157000	1.192600	1.193420
FEM-VCE	1.249100	1.221500	1.211280
FEM-QPE	1.158200	1.147200	1.147980
$\frac{K_{ref16}}{S\sqrt{\pi a}}$	1.2156		

Table 5: Values of the stress intensity factor; cracks emanating from a circular hole; $2R/W = 0.25$, $2a/W = 0.4$.

$\frac{K_I}{S\sqrt{\pi a}}$	mesh-A	mesh-B	mesh-C
BEM-SST	1.286054	1.286096	1.285191
FEM-JCI	1.153300	1.226100	1.226751
FEM-VCE	1.188400	1.205109	1.209227
FEM-QPE	1.204700	1.216600	1.217284
$\frac{K_{ref16}}{S\sqrt{\pi a}}$	1.2853		

Table 6: Values of the stress intensity factor; cracks emanating from a circular hole; $2R/W = 0.25$, $2a/W = 0.5$.

8 Discussion and Conclusions

The quarter-point finite element results appear to be the least accurate solution among all the different approaches tested. This lack of accuracy is due to the fact that the quarter-point finite element solution is dependent on the size of the quarter-point element. As a matter of fact, Ingraffea and Manu [17] and Lynn and Ingraffea [18] have shown that the size of the quarter-point elements affect the computed stress intensity factor in such a way that, all other details remaining constant, there is a particular crack tip element

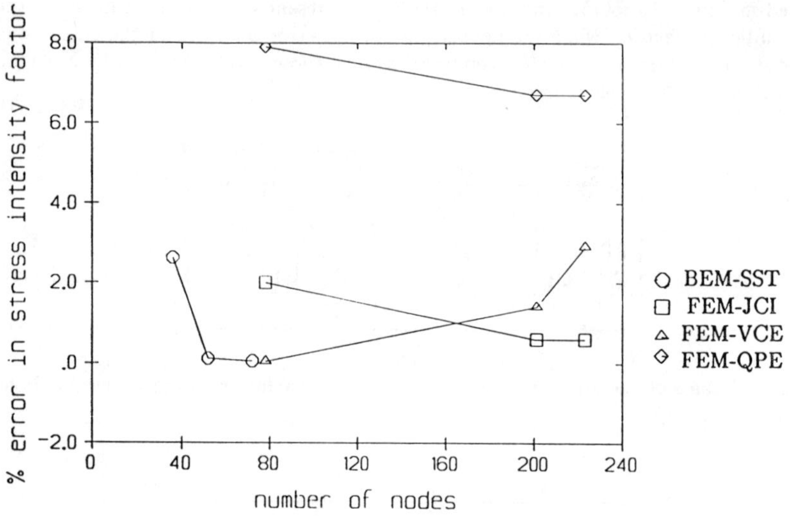

Figure 13: Error in the stress intensity factor calculation; cracks emanating from a circular hole; $2R/W = 0.25$, $2a/W = 0.3$.

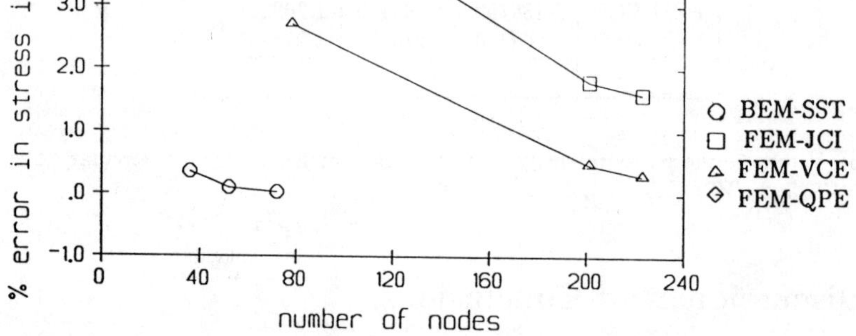

Figure 14: Error in the stress intensity factor calculation; cracks emanating from a circular hole; $2R/W = 0.25$, $2a/W = 0.4$.

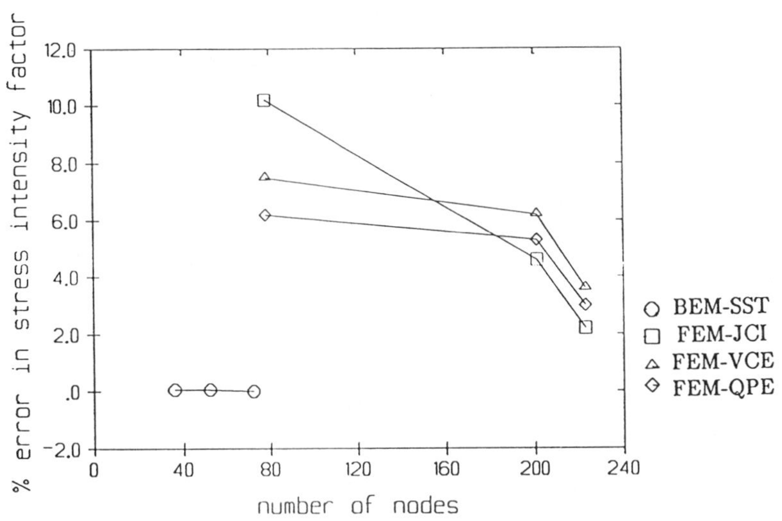

Figure 15: Error in the stress intensity factor calculation; cracks emanating from a circular hole; $2R/W = 0.25$, $2a/W = 0.4$.

size which minimizes the error in the computed stress intensity factor. The reasons for this size dependence are due to the need to simultaneously represent the singular and finite stress terms in a given problem, as it is explained by Harrop [19]. As a consequence of this dependence, it is clearly impossible to recommend a particular crack-tip element size, whenever the problem solution is not known *apriori*.

The J-integral finite element approach performs better than the quarter-point element. Nevertheless, it converges to values of the stress intensity factor which still present relatively high percentage of error.

The virtual crack-extension approach presented some of the most accurate values obtained for the stress intensity factor, among all the finite element techniques considered. However, the accuracy of the method depends not only on the amount of the virtual extension, Hellen [6], but also on the nodes released in the near tip region along with the virtual crack-extension. In order to check this source of inaccuracy, further calculations were carried out, keeping always the virtual crack-extension between the bounds recomended by Hellen [6]. The first test considered a virtual crack-extension of 0.001 and released the nodes of the two crack-tip closest elements. The second test considered a virtual extension of 0.0005 and released the nodes of the same elements of the first test. The third test released the nodes of the eight crack-tip closest elements for a virtual extension of 0.001. The results obtained are presented in table 7. These dependencies on the amount of the virtual extension and on the released nodes constitute the main pitfall of the technique.

Lastly, the singularity subtraction technique presents the best results for the stress intensity factor, regarding both the accuracy as well as the convergence. Furthermore, such high levels of accuracy were easily achieved with no more than a few elements. This is a consequence of the fact that the crack-tip singularity subtraction induces a smooth elastic field which is accurately solved by the boundary element method. Most importantly, the singularity subtraction technique is applied in an automatic way, with no dependence on any parameter.

$2a/W$	$\frac{K_{ref16}}{S\sqrt{\pi a}}$	case A		case B		case C	
		$\frac{K}{S\sqrt{\pi a}}$	% error	$\frac{K}{S\sqrt{\pi a}}$	% error	$\frac{K}{S\sqrt{\pi a}}$	% error
0.3	1.0583	1.00974	4.5	1.03006	2.6	1.05785	0.04
0.4	1.1102	1.08397	2.3	1.10387	0.5	1.06954	3.6
0.5	1.1876	1.14152	3.8	1.14586	3.5	1.08569	8.5

Table 7: Stress intensity factors evaluated by the virtual crack-extension finite element approach; central crack problem; case A - 2 elements released and virtual extension=0.001; case B - 2 elements released and virtual extension=0.0005; case C - 8 elements released and virtual extension=0.001

References

[1] Henshell, R.D. and Shaw, K.G., "Crack Tip Elements are Unnecessary", *Int. J. for Num. Meth. Engng.*, 9, 495-509, 1975.

[2] Barsoum, R. S., "On the Use of Isoparametric finite elements in Linear Fracture Mechanics", *Int. J. Num. Meth. Engng.* 10, No.1, 25-37, 1976

[3] Rice, J.R., "A Path-independent Integral and the Approximate Analysis of Strain Concentration by Notches and Cracks", *Trans. ASME J. Appl. Mech.*, 35,379-386, 1968.

[4] Griffiths, A. A., *Phil. Trans. Roy. Soc. London*, Ser. A.221, 163, 1921.

[5] Parks, D. M., "A Stiffness Derivative Finite Element Technique for Determination of Elastic Crack Tip Stress Intensity Factors", *Int. J. Fracture*, 10, No.4, 487-502, 1974.

[6] Hellen, T.K., "On the Method of Virtual Crack Extensions", *Int. J. Num. Meth. Engng.*, 9, No.1, 187-208, 1975.

[7] Pian, T. H. H., "Derivation of Element Stiffness Matrices for Assumed Stress Distributions", *J.A.I.A.A.*, 2, No.7, 1333, July 1964.

[8] Martinez, J., Dominguez, J., "On the Use of Quarter Point Boundary Elements for Stress Intensity Factor Computations", *Int. J. Num. Meth. Engng.*, 20, 1941-1950.

[9] Cruse, T. A., "Two Dimensional BIE Fracture Mechanics Analysis", *Appl. Math. Modl.*, 2, 287-293.

[10] Crouch, S. L., "Solution of Plane Elasticity Problems by the Displacement Discontinuity Method", *Int. J. Num. Meth. Engng.*, 10, 301-343, 1976.

[11] Aliabadi, M.H., Rooke, D.P., Cartwright, D.J., "An Improved Boundary Element Formulation for Calculating Stress Intensity Factors: Application to Aerospace Structures", *J. Strain Anal.*, 22 (4), 1–5, 1987.

[12] Gallagher, R. H., "A Review of Finite Element Techniques in Fracture Mechanics", Proc. 1st Int. Conf. on Num. Meth. in Fracture Mechanics, A. R. Luxmoore and D. R. J. Owen, Eds. Swansea, 9-13, January 1978.

[13] Smith, R. N. L., "From Rags to Riches?-Developments in the BEM for the Solution of Elastic Fracture Problems", Cranfield Institute of Technology, Computational Mathematics Group, Technical Note COMAG 1/88, 1988.

[14] Williams, M.L., "Stress Singularities Resulting from Various Boundary Conditions in Angular Corners of Plates in Extension", *J. Appl. Mech.*, 526–528, 1952.

[15] Smith, R. N. L., Aliabadi, M. H., "Boundary Integral Equation Methods for the Solution of Crack Problems", *Computers and Mathematics with Applications*, Special issue on BEM, to appear.

[16] Newman, J. C., "An Improved Method of Collocation for Stress Analysis of Cracked Plates with Various Shaped Boundaries", NASA Langley Research Center, Report NASA TND-6376, 1971.

[17] Ingraffea, A. R. and Manu, C., "Stress-intensity Factor Computation in Three Dimensions with Quarter-point Elements", *Int. J. Num. Meth. Engng.*, 15, 1427-1445, 1980.

[18] Lynn, P. P. and Ingraffea, A. R., "Transition Elements to be Used with Quarter-point Crack-tip Elements", *Int. J. Num. Meth. Engng.*, 12, 1031-1036, 1978.

[19] Harrop, L. P., "The Optimum Size of Quarter-Point Crack Tip Elements", *Int. J. for Num. Meth. Engng.*, 1101-1103, 1982.

Multilevel Solution Methods and Automatic Node Moving for the Boundary Element Method

W.S. Hall and T.T. Hibbs

Division of Mathematics and Statistics, School of Information Engineering, Teesside Polytechnic, Middlesbrough, TS1 3BA, Cleveland, UK

ABSTRACT

The paper describes work which is aimed at providing efficient, automatic fracture calculations using the Boundary Element Method. Hierarchical shape functions are used to define a simple estimated solution error which is the basis of a node moving calculation applied to two integral equations. Hierarchical shape functions also provide a natural definition of different levels of approximation for a multilevel solution procedure and it is seen that savings in calculation time would result from the common use of some solution approximations in the moving node and multilevel procedures.

INTRODUCTION

The study of cracks in structures is of wide interest, particularly for aircraft. The work described in this paper is aimed at improving computer analysis of cracks by creating efficient, automatic fracture calculations using the Boundary Element Method.

The stress field at the tip of a crack is difficult to represent accurately using normal numerical calculation procedures and a method of producing a suitable calculation mesh is needed. A basis is presented for creating a graded mesh automatically, using moving nodes associated with an efficient multilevel solution procedure.

Fundamental ideas concerning hierarchical shape functions, multilevel solution methods and moving nodes are described, together with applications to one dimensional integral equations. These ideas will be extended in due course to the two dimensional, and even three dimensional, equations of elastostatics governing fracture mechanics.

Adaptive methods have been pioneered in the Finite Element Method. The simplest of these is the so called h-adaptive method in which additional nodes are introduced into the mesh in areas where the behaviour of the solution requires them. An alternative adaption procedure keeps the defining nodes for the solution unchanged and changes the level of the polynomial approximation to the solution between them. It is termed p-adaptive because the level of the pth order polynomials is varied.

A third adaptive scheme is the r-adaptive method, or redistributive adaption, in which nodes are moved from their original position, for example to bunch together in areas of high error. This is also called moving node adaption.

Hierarchical shape functions developed for the Finite Element Method by Zienkiewicz, Gago and Kelly (1983) have recently been connected to multilevel techniques by Zienkiewicz and Craig (1986) and to p-adaptive methods by Zhu and Zienkiewicz (1987). Hierarchical shape functions also form the basis of recent work by Alarcon and Reverter (1987) on p-adaptive methods for the Boundary Element Method. In the work aimed at Boundary Element Methods described in this paper the connection between multilevel techniques and hierarchical shape functions is again used, though moving node adaptive techniques are substituted for p-adaptive methods.

HIERARCHICAL SHAPE FUNCTIONS

In a hierarchical shape function approximation, a function f(x) is expanded over the interval [-1,1] as

$$f(x) = N_0(x)f_{-1} + N_1(x)f_1 + N_2(x)f_0^{(2)} + N_3(x)f_0^{(3)} + N_4(x)f_0^{(4)} \quad (1)$$

where f_{-1} and f_1 are nodal values at the ends x = -1 and x = 1 of the interval, or element, and $f_0^{(2)}$, $f_0^{(3)}$ and $f_0^{(4)}$ are second, third and fourth derivatives evaluated at the centre x = 0. In equation (1) the form of the shape functions is as follows

$$N_0(x) = (1-x)/2 \ , \quad N_1(x) = (1+x)/2 \ , \quad N_2(x) = (x^2-1)/2 \ ,$$

$$N_3(x) = (x^3-x)/6 \ , \quad N_4(x) = (x^4-1)/24 \quad (2)$$

where N_0, N_1 are linear shape functions and N_2, N_3 and N_4 are respectively quadratic, cubic and quartic bubble functions.

These are the shape functions developed by Peano (1975), although other families are available. They have the advantage, starting from the linear approximation containing only two terms, of adding extra terms without altering earlier calculations. This has been demonstrated in Hall and Hibbs (1988) applied to an ordinary integral equation

$$f(x) = g(x) + \int_{-1}^{1} K(x,y)f(y)dy \ . \tag{3}$$

Substituting f(x) given by equation (1) and with collocation points at x = -1, +1, 0, -0.5 and +0.5 leads to four sets of discretized equations of the form

$$\underline{K}_i\underline{F}_i = \underline{g}_i \qquad i = 1,\ 2,\ 3,\ 4 \tag{4}$$

each corresponding to a level of approximation, linear to quartic. As is shown in Hall and Hibbs (1988) the coefficient matrices at the different levels are of the form

$$\underline{K}_1 = \begin{bmatrix} \underline{k}_{11} \end{bmatrix}, \qquad \underline{K}_2 = \begin{bmatrix} \underline{k}_{11} & \underline{k}_{12} \\ \underline{k}_{21} & \underline{k}_{22} \end{bmatrix}$$

$$\underline{K}_3 = \begin{bmatrix} \underline{k}_{11} & \underline{k}_{12} & \underline{k}_{13} \\ \underline{k}_{21} & \underline{k}_{22} & \underline{k}_{23} \\ \underline{k}_{31} & \underline{k}_{32} & \underline{k}_{33} \end{bmatrix}, \qquad \underline{K}_4 = \begin{bmatrix} \underline{k}_{11} & \underline{k}_{21} & \underline{k}_{23} & \underline{k}_{24} \\ \underline{k}_{21} & \underline{k}_{22} & \underline{k}_{23} & \underline{k}_{24} \\ \underline{k}_{31} & \underline{k}_{32} & \underline{k}_{33} & \underline{k}_{34} \\ \underline{k}_{41} & \underline{k}_{42} & \underline{k}_{43} & \underline{k}_{44} \end{bmatrix} \tag{5}$$

in which an additional row and column are added as the level increases. Also

$$\underline{F}_4 = \begin{bmatrix} f_{-1},\ f_1,\ f_0^{(2)},\ f_0^{(3)},\ f_0^{(4)} \end{bmatrix}^T \tag{6}$$

and

$$\underline{g}_4 = \begin{bmatrix} g_{-1},\ g_1,\ g_0,\ g_{-0.5},\ g_{0.5} \end{bmatrix}^T \tag{7}$$

where $[\ ]^T$ indicates transpose. Going down from the highest, quartic level given by $\underline{F}_4$ and $\underline{g}_4$ is achieved by removing elements from $\underline{F}_4$ and $\underline{g}_4$ until

$$\underline{F}_1 = \begin{bmatrix} f_{-1},\ f_1 \end{bmatrix}^T, \qquad \underline{g}_1 = \begin{bmatrix} g_{-1},\ g_1 \end{bmatrix}^T \tag{8}$$

are reached.

It can be seen that $\underline{F}_i$ is not the solution in terms of nodal values. Such a solution, however, may be obtained from

$$\underline{f}_i = \underline{B}_i\underline{F}_i \quad , \tag{9}$$

where

$$\underline{f}_4 = \left[f_{-1}, f_1, f_0, f_{-0.5}, f_{0.5} \right]^T . \tag{10}$$

The lower level solutions $\underline{f}_3$, $\underline{f}_2$ and $\underline{f}_1$ are defined by removing the last element successively. Hall and Hibbs (1988) give the form of the matrix $\underline{B}_i$.

EQUATION RESIDUAL AND ESTIMATED SOLUTION ERROR

The different levels of problem discretisation coming from the use of hierarchical shape functions lead to the definition of an error in the solution corresponding to each level. On the other hand it is a feature of multilevel techniques that being iterative methods of solution, they also involve errors though these are equation residuals, $\underline{r}$, which are defined by

$$\underline{r} = \underline{g} - \underline{Kf}^{app} \tag{11}$$

for an equation

$$\underline{Kf} = \underline{g} \tag{12}$$

where $\underline{K}$ and $\underline{g}$ are given and $\underline{f}$ is unknown. The term $\underline{f}^{app}$ is an approximation to the unknown after some stage of the iteration. The residual $\underline{r}$ is clearly a measure of the amount by which the equation is not satisfied at different points and it may be thought that improvements made to the level of approximations where $\underline{r}$ is largest would reduce its value.

Returning to the solution error, the difference

$$\underline{d} = \underline{f} - \underline{f}^{app} \tag{13}$$

between the solution $\underline{f}$ and some approximation to it, $\underline{f}^{app}$, measures directly the amount by which the solution itself is not achieved. If $\underline{d}$ was known exactly then the exact solution could be obtained immediately from

$$\underline{f} = \underline{f}^{app} + \underline{d} \quad . \tag{14}$$

In the course of a calculation, $\underline{d}$ would not be known. However some estimate $\underline{d}_e$ of it might be known and this could be used to make local improvements to the level of approximation of the problem in places where $\underline{d}_e$ was largest. Thus a distinction must be made between the equation residual

$$\underline{r} = \underline{g} - \underline{Kf}^{app} \tag{11}$$

and the estimated solution error

$$\underline{d}_e - (\underline{f} - \underline{f}^{app})_e \quad , \tag{15}$$

although both can be used to drive an adaptive process.

ESTIMATED SOLUTION ERRORS OVER AN ELEMENT

The three types of adaptive methods (h, p and r) have been outlined in the Introduction. Also the point values of the equation residual and the estimated solution error have been outlined in the previous section. However, in an application of adaptive methods to the Boundary Element Method, the controlling equation residual or estimated solution error will be required over each element, rather than at a point. The definition of the element solution error and ways of estimating it are now given.

The interval of an integral equation, say [a,b], is divided into subintervals, i, as shown in Figure 1. In the extension to the curved boundary of a Boundary Integral Equation these would, of course, be described as elements.

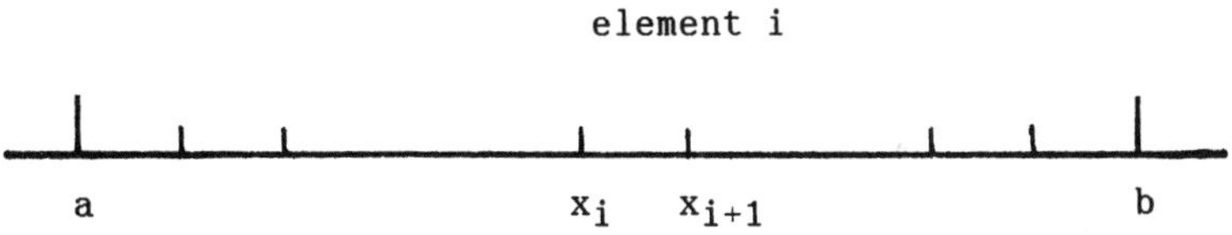

Figure 1. Division of integral equation range [a,b] into subintervals.

The solution error for element i is defined as

$$e_i = \int_{x_i}^{x_{i+1}} [\underline{f}(x) - \underline{f}^{app}(x)]^2 dx \ . \tag{16}$$

In the above definition the exact solution, $\underline{f}(x)$, must itself be estimated by some other quantity. A possibility is simply to take the next highest approximation to $\underline{f}^{app}$. For example, it would be possible to take linear and quadratic hierarchical approximations so that

$$\underline{f}^{app} = \underline{f}_1 \quad , \qquad \underline{f} = \underline{f}_2 \quad . \tag{17}$$

Or, at a higher level of approximation involving quadratic and cubic approximations,

$$\underline{f}^{app} = \underline{f}_2 \quad , \qquad \underline{f} = \underline{f}_3 \quad . \tag{18}$$

In this paper only the lower level of approximation will be considered.

The element $[x_i, x_{i+1}]$ is mapped to the standard interval $[-1,1]$ described by the parameter u using

$$x = \frac{1}{2}(x_i+x_{i+1}) + \frac{1}{2}u(x_{i+1}-x_i)$$

$$= \bar{x}_i + \frac{1}{2}ud_i \tag{19}$$

where $\bar{x}_i$ is the mid position and d_i the element length. Thus

$$e_i = \int_{-1}^{1} [\underline{F}(u)-\underline{F}^{app}(u)]^2 \frac{1}{2} d_i du \quad , \tag{20}$$

where

$$\underline{F}(u) = \underline{f}(x(u)) \quad , \quad \underline{F}^{app}(u) = \underline{f}^{app}(x(u)) \quad . \tag{21}$$

Taking the linear and quadratic hierarchical approximations for $\underline{F}^{app}(u)$ and $\underline{F}(u)$ gives

$$\underline{F}^{app}(u) \equiv \underline{F}_1(u) = N_0(u)\underline{F}_{-1} + N_1(u)\underline{F}_1 \quad , \tag{22}$$

$$\underline{F}(u) \equiv \underline{F}_2(u) = N_0(u)\underline{F}_{-1} + N_1(u)\underline{F}_1 + N_2(u)\underline{F}_0^{(2)} \quad . \tag{23}$$

Thus for that estimated solution error which is based on the difference between the linear and quadratic approximations,

$$e_i = \int_{-1}^{1} [N_2(u)\underline{F}_0^{(2)}]^2 \frac{1}{2} d_i du = \frac{d_i}{2}[\underline{F}_0^{(2)}]^2 \int_{-1}^{1} \frac{1}{4}(u^2-1)^2 du$$

$$= \frac{2}{15} d_i [\underline{F}_0^{(2)}]^2$$

$$= \frac{d_i^5}{120}[\underline{f}^{(2)}(\bar{x})]^2 \quad , \tag{24}$$

since

$$[\underline{F}_0^{(2)}]^2 = \frac{d_i^4}{16}[\underline{f}^{(2)}(\bar{x})]^2 \quad . \tag{25}$$

The above formula for e_i has the important advantage that it reduces an integration to a single evaluation involving firstly a simple geometric quantity d_i, the element length, and secondly $f^{(2)}(\bar{x})$, which comes as part of the hierarchical solution.

MOVING NODE ADAPTION BASED ON ESTIMATED SOLUTION ERRORS

The r-adaptive, or moving node scheme, requires the nodes to be moved to make e_i the same for all elements i. For a moving node scheme adaption occurs at every stage, in contrast to h-adaptive or p-adaptive schemes where some threshold must be passed in order to introduce a new node or to increase the level of approximation from p to p+1. The amount of movement of each node will be based on the element solution error given by equation (24). The process by which the movements are determined has been performed according to the Cook and Duncan algorithm (1982) as reported in Hall and Hibbs (1988) and here by a more direct algorithm given in Hall and Hibbs (1989).

Although this work is aimed at boundary integral equations first results have been found for two integral equations. Thus the investigation deals with a single central moving node at X in the interval [-1,1] of the equations below

$$f(x) = \frac{1}{24}\int_{-1}^{1} (1+x)(1+y)f(y)dy + \cosh\left[\frac{1}{2}(1+x)\right] - \frac{1}{4}(1+x) , \quad (26)$$

$$f(x) = -\frac{1}{4}\int_{-1}^{1} \left[(1+x)(1+y)\right]^{1/2} f(y)dy + \left[\frac{(1+x)}{2}\right]^{1/2} . \quad (27)$$

The calculations carried out use automatic node moving of the single node based on an estimated element solution error given by

$$e_i = \int_{x_i}^{x_{i+1}} \left[f(x) - f^{app}(x) \right]^2 dx \quad , \quad i = 1,2 \quad (28)$$

to equalise the errors e_1 and e_2. In equation (28), f(x) has been taken as $f_2(x)$, the quadratic approximation, and $f^{app}(x)$ as $f_1(x)$, the linear approximation.

The results of some numerical experiments are now presented. Results for integral equation (26) are given in Table 1 and results for integral equation (27) are given in Table 2. The entries in the tables are the position of the central moving node for each iteration number.

STARTING POINT / ITERATION NUMBER	0.0	0.1	0.2	0.3
1	0.04470	0.04702	0.05712	0.08864
2	0.04614	0.04614	0.04629	0.04674
3			0.04614	0.04614

Table 1. Moving node results for integral equation (26).

STARTING POINT / ITERATION NUMBER	-0.6	-0.5	-0.4	-0.3
1	-0.5870	-0.5468	-0.4899	-0.4230
2	-0.5830	-0.5682	-0.5419	-0.5041
3	-0.5817	-0.5765	-0.5660	-0.5488
4	-0.5812	-0.5795	-0.5757	-0.5690
5	-0.5809	-0.5805	-0.5792	-0.5768
6		-0.5808	-0.5804	-0.5796
7		-0.5809	-0.5808	-0.5805
8			-0.5809	-0.5808
9				-0.5809

Table 2. Node moving results for integral equation (27).

It can be seen that, for both integral equations and all starting values, the node moves towards its converged value from one side only in a stable way in few steps. More steps are required for equation (27) because the solution has an infinite slope at $x = -1$.

MULTIGRID/MULTILEVEL METHODS

Multigrid or multilevel methods are a class of iterative solution techniques for solving the discretized equations arising from a wide variety of sources. Multigridding uses a first, finest grid discretization together with successive coarser discretizations as shown in Figure 2.

Usual iterative solution methods quickly remove Fourier components of the error whose wavelength is that of the grid spacing, but long wavelengths are removed slowly. Multigrid methods, however, also quickly remove error components with long wavelengths by working at larger grid sizes.

Multilevel methods are a generalisation of multigrid methods, and include the latter as a special case. They are formulated in terms of different levels of approximation of the basic equations rather than different grid sizes. The levels may be different polynomial approximations to the unknown functions (eg. hierarchical shape functions) though, of course, the different grid sizes previously referred to may also be considered as different levels of approximation.

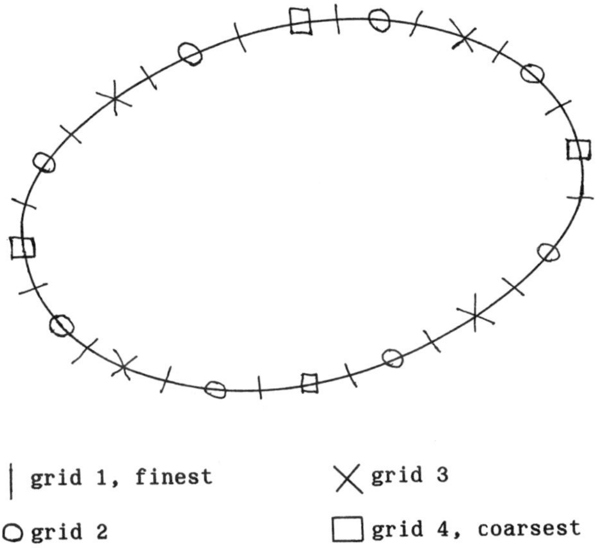

Figure 2. Typical four grid approximation.

With hierarchical shape functions the levels are given by the linear, quadratic, cubic and quartic representations. Hierarchical shape functions provide natural representations for multilevel methods and the contractions and prolongations between levels are straightforward.

The hierarchical coefficients of equation (5) form the basis of the calculation

linear	$\underline{k}_{11}$	$\underline{k}_{12}$	$\underline{k}_{13}$	$\underline{k}_{14}$	,
quadratic	$\underline{k}_{21}$	$\underline{k}_{22}$	$\underline{k}_{23}$	$\underline{k}_{24}$	,
cubic	$\underline{k}_{31}$	$\underline{k}_{32}$	$\underline{k}_{33}$	$\underline{k}_{34}$	,
quartic	$\underline{k}_{41}$	$\underline{k}_{42}$	$\underline{k}_{43}$	$\underline{k}_{44}$	.

The sets of equations are thus given by

$$\underline{K}_i\underline{F}_i = \underline{g}_i \text{ for } i = 1,2,3,4 \quad . \qquad (29)$$

<u>Downward part of multilevel cycle</u>

The part of the multilevel cycle going from higher to lower levels is now described. At level 4 (the quartic level), the equations are

$$\underline{K}_4\underline{F}_4 = \underline{g}_4 \quad , \tag{30}$$

which can be solved iteratively to find $\underline{F}_4^{app}$, the first approximation, and the residual calculated from

$$\underline{r}_4 = \underline{g}_4 - \underline{K}_4\underline{F}_4^{app} \quad , \tag{31}$$

where $\underline{r}_4$ has components $\underline{r}_{41}$, $\underline{r}_{42}$, $\underline{r}_{43}$, $\underline{r}_{44}$ corresponding to the different levels of approximation 1 to 4 given by

$$\underline{r}_4 = \begin{bmatrix} \underline{r}_{41} & \underline{r}_{42} & \underline{r}_{43} & \underline{r}_{44} \end{bmatrix}^T . \tag{32}$$

At level 3 (the cubic level), $\underline{r}_4$ is contracted to $\underline{d}_3$ by leaving off the last component $\underline{r}_{44}$ corresponding to the quartic level. Therefore

$$\underline{d}_3 = D(\underline{r}_4) = \begin{bmatrix} \underline{r}_{41} & \underline{r}_{42} & \underline{r}_{43} \end{bmatrix}^T , \tag{33}$$

where D indicates the contraction ('down') operator. The equation

$$\underline{K}_3\underline{v}_3 = \underline{d}_3 \tag{34}$$

is solved iteratively for $\underline{v}_3^{app}$ and then the residual

$$\underline{r}_3 = \underline{d}_3 - \underline{K}_3\underline{v}_3^{app} \tag{35}$$

is calculated.

The same process continues to level 2 and level 1 with the sole exception that the level 1 equations are solved directly by elimination. Thus correction vectors $\underline{v}_1$, $\underline{v}_2^{app}$ and $\underline{v}_3^{app}$ to the approximate solution $\underline{F}_4^{app}$ are produced which will be incorporated in an upward cycle, a description of which follows.

The upward part of the multilevel cycle

The part of the multilevel cycle going from lower to higher levels is now described.

At level 2 (quadratic) it is required to correct $\underline{v}_2^{app}$ using $\underline{v}_1$. However, an element is missing and a prolongation operator is required to provide it. Consider the equations

$$\underline{K}_2\underline{v}_2 = \underline{d}_2 \tag{36}$$

which may be written more fully as

$$\begin{bmatrix} \underline{k}_{11} & \underline{k}_{12} \\ \underline{k}_{21} & \underline{k}_{22} \end{bmatrix} \begin{bmatrix} \underline{v}_1 \\ \underline{v}_2^e \end{bmatrix} = \begin{bmatrix} \underline{r}_{31} \\ \underline{r}_{32} \end{bmatrix} \tag{37}$$

where the $\underline{k}$'s are known and $\underline{v}_1$, $\underline{r}_{31}$ and $\underline{r}_{32}$ are also known. The second component $\underline{v}_2^e$ is the component corresponding to the quadratic level which is unknown but may be estimated by writing the second element of the matrix equation

$$\underline{k}_{21}\underline{v}_1 + \underline{k}_{22}\underline{v}_2^e = \underline{r}_{32} \quad . \tag{38}$$

Therefore

$$\underline{v}_2^e = \underline{k}_{22}^{-1}[\underline{r}_{32} - \underline{k}_{21}\underline{v}_1] \quad . \tag{39}$$

The prolongation of $\underline{v}_1$, called $\underline{c}_2$ ie. $\underline{c}_2 = P(\underline{v}_1)$, is thus given by

$$\underline{c}_2 = \begin{bmatrix} \underline{v}_1 & \underline{v}_2^e \end{bmatrix}^T \quad . \tag{40}$$

Therefore

$$\underline{v}_2 = \underline{v}_2^{app} + \underline{c}_2 \quad . \tag{41}$$

The value of $\underline{v}_2$ found above is used as the first approximation in an iterative solution of

$$\underline{K}_2\underline{v}_2 = \underline{d}_2 \tag{42}$$

to obtain an improved value of $\underline{v}_2$.

At level 3 (cubic) the required prolongation of $\underline{v}_2$ is $\underline{c}_3$, which is given by a similar expression to (39), where

$$\underline{c}_3 = P(\underline{v}_2) \quad . \tag{43}$$

Therefore

$$\underline{v}_3 = \underline{v}_3^{app} + \underline{c}_3 \quad . \tag{44}$$

Again the value of $\underline{v}_3$ above is used as the first approximation in the iterative solution of

$$\underline{K}_3\underline{v}_3 = \underline{d}_3 \tag{45}$$

to obtain an improved solution of $\underline{v}_3$.

At level 4 (quartic) the required prolongation of $\underline{v}_3$ is $\underline{c}_4$, again given by a similar expression to (39) where

$$\underline{c}_4 = P(\underline{v}_3) \quad . \tag{46}$$

Therefore

$$\underline{F}_4 = \underline{F}_4^{app} + \underline{c}_4 \quad . \tag{47}$$

The value of $\underline{F}_4$ is used in the iterative solution of

$$\underline{K}_4\underline{F}_4 = \underline{g}_4 \tag{48}$$

to give the final solution in terms of derivative quantities and

$$\underline{f}_4 = \underline{B}_4\underline{F}_4 \tag{49}$$

gives the nodal values of the solution.

The procedure described in which the iterations are performed at the highest level proceeding down to the lowest level and back to the highest level is known as one complete V cycle. The solution procedure may consist of several V cycles.

There is a clear link between:-

(a) Hierarchical shape functions and moving nodes, since hierarchical shape functions provide a simple evaluation of an estimated solution error over an element which may be used to control the node movement.

(b) Hierarchical shape functions and the multilevel method, since they provide a natural definition of the different approximation levels with a simple means of carrying out prolongations and contractions.

Hierarchical shape functions thus provide an indirect link between moving node calculations and the multilevel method of solution. There is also a direct link as indicated in the outline moving node and solution algorithm which follows.

1. Specify a set of initial values of nodal positions X_i.
2. Calculate $\underline{f}_1$ and $\underline{f}_2$ using the multilevel procedure.
3. Modify the nodal positions, X_i, and then return to step 2, continuing until convergence occurs.
4. Expand the multilevel description of the problem based on the final nodal positions which entails creating the matrices $\underline{K}_3$ and $\underline{K}_4$.
5. Perform a four level multilevel solution using the matrices $\underline{K}_1$, $\underline{K}_2$, $\underline{K}_3$ and $\underline{K}_4$.

CONCLUSIONS

It has been shown that a link exists between hierarchical shape functions, automatic node moving and multilevel methods. The use of hierarchical shape functions together with a moving node scheme leads to stable convergence of nodal positions for tests on two ordinary integral equations. Some of the calculations carried out in the moving node process can also be used in a multilevel solution procedure.

This work has been carried out with the support of Procurement Executive Ministry of Defence.

REFERENCES

E.Alarcon and E.Reverter (1986), 'P-Adaptive Boundary Elements', Int. J. Numer. Meths. Engng., Vol.23, pp 801-829.

A.Brandt (1977), 'Multi-level Adaptive Solutions to Boundary Value Problems', Math. Comp., Vol.31, pp 333-390.

A.E.Cook and R.Duncan (1982), 'A Node Moving Algorithm with Application to Burgers Equation and the Moltz Problem', Appl. Math. Mod., Vol.6, pp 463-468.

W.S.Hall and T.T.Hibbs (1988), 'Hierarchical Shape Functions, Multilevel Solution Methods and Moving Nodes for the Boundary Element Method', Teesside Polytechnic Maths. Report TPMR 88-6.

W.S.Hall and T.T.Hibbs (1989), 'Crack Tip Hierarchical Shape Functions and a Direct Node Moving Algorithm', Teesside Polytechnic Maths. Report TPMR 89-3.

P.Parriera (1987), 'Self Adaptive P-Hierarchical Boundary Elements in Elastostatics', Boundary Elements IX, Editors C.A.Brebbia, W.L.Wendland and G.Kuhn, Springer-Verlag.

A.G.Peano (1975), 'Hierarchies of Conforming Finite Elements', Doctural dissert., Washington Univ., St. Louis.

O.C.Zienkiewicz and A.Craig (1986), 'Adaptive Refinement, Error Estimates, Multigrid Solution and Hierarchic Finite Element Method Concepts', Chapter 2 of 'Accuracy Estimates and Adaptive Refinements in Finite Element Computations' edited by I.Babuska, O.C.Zienkiewicz, J.Gago and E.R. de A.Oliveira, John Wiley.

O.C.Zienkiewicz, J.P. de S.R.Gago and D.W.Kelly (1983), 'The Hierarchical Concept in Finite Element Analysis', Comps. and Structs., Vol.16, pp 63-65.

J.Z.Zhu and O.C.Zienkiewicz (1988), 'Adaptive Techniques in the Finite Element Method', Commun. Appl. Numer. Meths., Vol.4, pp 197-204.

Numerical Error Estimates for the Three-Dimensional Boundary Element Method

S.H. Crook and R.N.L. Smith

Applied and Computational Mathematics Group, RMCS, Shrivenham, Swindon, SN6 8LA, Wiltshire, UK

ABSTRACT

Error estimates may be derived from the residual approach for three-dimensional elasticity problems. Estimates of residuals have previously been obtained mainly in two dimensions using analytical integration. For more general cases numerical integration is necessary. Residual estimates are difficult to obtain with numerical integration procedures as we show that these may lead to severe oscillations in the isoparametric form. An integration scheme which leads to an adequately smooth representation of the residual is described and the corresponding error estimates are then applied to problems with known solutions showing a useful correlation with the actual error.

INTRODUCTION

The Boundary Element Method is now established as a reliable and powerful tool for the solution of linear elliptic problems. Much interest now centres on the problem of obtaining an estimate of the error in values derived from a given mesh. These error estimates may then be used to change the mesh in a systematic fashion to reduce the error to an acceptable level - adaptive meshing.

There are two fundamental approaches to the derivation of error estimates. In the first approach, the difference between appropriate integrals of the applied and solution boundary values is calculated at points between interpolation nodes - the residual r. Error estimates are then computed based on suitable integral norms which are expressed in terms of r. In the second class of methods, a smooth solution is obtained by fitting a simple higher-order curve to the computed solution, and error estimates are found from measures of the goodness of fit or from fundamental properties of the higher order curve.

The three commonly implemented methods for adaptive meshing are, r-adaptive where refinement is achieved by rearrangement of existing nodes, p-adaptive where the polynomial degree of interpolation is increased, and h-adaptive where elements are subdivided. A fourth hybrid method, hp-adaptive, seeks to take advantage of the merits of h- and p-adaptive methods. An adaptive h-version removes the influence of singularities in the exact solution on the rate of convergence of the approximation. A p-version converges exponentially if the exact solution is smooth and always converges faster than a uniform refinement with fixed polynomial degree. Rank [7] has reported good results using the hp-version; it may be possible to achieve the same accuracy for a problem with a singularity by combining a subtraction of singularity technique as developed by Aliabadi [1] - making the solution smooth - and the p-version. All methods have to address the problem of error analysis, which requires calculation of error estimators (of the global error) and error indicators (of the local error) via residuals. The calculation of residuals is the fundamental step and is therefore considered in some detail here.

ERROR ESTIMATION

The direct formulation of the boundary element technique in the case of elasticity problems as given by Cruse [2] leads to equations of the form,

$$c_{ij}(\xi)u_j(\xi) + \int_\Gamma p^*_{ij}(\xi, \mathbf{x})u_j(\xi)d\Gamma(\xi) = \int_\Gamma u^*_{ij}(\xi, \mathbf{x})p_j(\xi)d\Gamma(\xi) \tag{1}$$

where Γ is the boundary, ξ is the source point, $\mathbf{x}$ is a field point, c_{ij} is dependent on the geometry at the source point ($c_{ij} = -\delta_{ij}/2$ if the surface is smooth), u_j, p_j represent displacement and traction components and u^*_{ij}, p^*_{ij} are the fundamental displacement and traction solutions. Equation (1) may be written in matrix form and the resulting linear matrix equation solved by an appropriate method. For a collocation solution to a BEM problem the approximating function is fitted exactly at all nodes on the mesh. That is,

$$Hu = Gt$$

is satisfied exactly so that $Hu - Gt = 0$. For a non-mesh point this solution will have a degree of error associated with it. Thus,

$$Hu - Gt = R$$

where R is known as the residual vector. The calculation of this vector is of primary concern, as it can be used to provide an error estimate for the approximate solution. When calculated locally to an element on the mesh it may also be used to provide an error indicator. Both error estimators and indicators are usually based on an integral norm of the residual function.

Error Indicators are defined locally to an element and are used to decide if local mesh refinement is required by providing an indication of the error in the solution. They are defined in terms of residuals calculated at points internal to the element. Parreira [6] and Rank [7] define error indicators for adaptive methods. The error indicator used in this work is defined as,

$$\chi_i = \left[\int_{\Gamma_i} \rho_i^2(\mathbf{z})d\Gamma(\mathbf{z})\right]^{1/2} \tag{2}$$

where $\rho_i(\mathbf{z})$ represents the variation of the residual R over element i.

Error Estimators provide an indication of the global error in the solution, and are used to determine whether further refinement of the mesh is necessary. An error estimator η satisfies the condition,

$$C_1\eta \leq ||e|| \leq C_2\eta \tag{3}$$

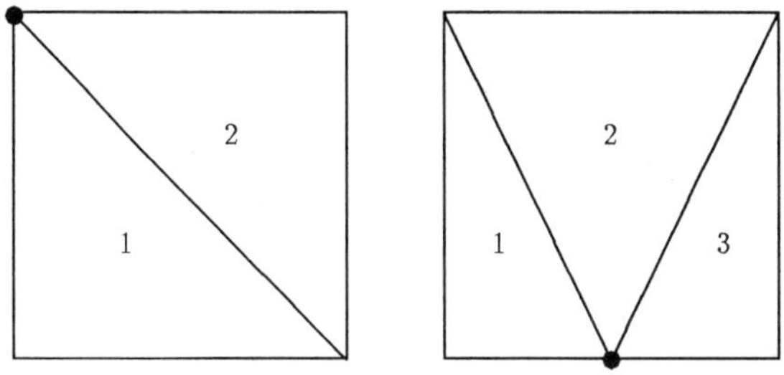

Figure 1: Partitions of Elements for Singular Integration

where $e = U - u$, u is the exact solution, U is the approximate solution, and for constants C_1, C_2 independent of u. Unfortunately C_1, C_2 cannot be determined, but a similar result to Equation (3) has been observed numerically for a large class of problems in the FEM, and is assumed to hold for the BEM. The error estimator is defined by,

$$\eta^2 = \sum_{i=1}^{N_e} \int_{\Gamma_i} \rho^2(\mathbf{z}) d\Gamma(\mathbf{z}) = \sum_{i=1}^{N_e} \chi_i^2.$$

Whilst it is important to estimate the integral of the residual function to some known accuracy, attention must be paid to the cost of such an approximation, as evaluating the residual at one point requires integration over the whole boundary. Since the error estimator is used to determine whether the refinement is necessary it need not be determined as accurately as the final BEM solution. An error in the estimate will cause unnecessary refinement or fail to implement required refinement which is inefficient but does not introduce any additional error into the BEM solution. The integration method used is described below, and a series of results validating this approach are given.

SINGULAR INTEGRATION

The calculation of residual values involves further numerical integration after finding the solution of the original problem. Each of the residual nodes is considered as a source point for integrals over all elements in the solution sub-region. Particular difficulty occurs when considering integration over an element containing the source point. These integrals are singular and the standard Gaussian integration technique becomes uneconomic to use.

A reliable method for the evaluation of the singular integrals on an isoparametric element is given by Lachat and Watson [5] and is used as standard throughout our work. The element is subdivided into triangular partitions and each partition is mapped onto a quadrilateral with the source point being expanded onto one side. A standard Gaussian integration rule may be used to evaluate the integrals on these quadrilaterals giving the element contribution as the sum over each of them. The number of sub-divisions necessary to evaluate the singular integrals is dependent on whether the collocation point is a vertex-node or a mid-edge mode (Figure (1)). When the source point is internal to the element - as for residual points - a similar approach to Lachat and Watson may be used. The element is divided into four partitions as shown in Figure (2), each is transformed as before and the integral calculated as the sum of the contributions over each partition.

A second approach for the evaluation of the singular integrals is to divide the element into four sub-elements introducing temporary nodes as required. The source point is at a

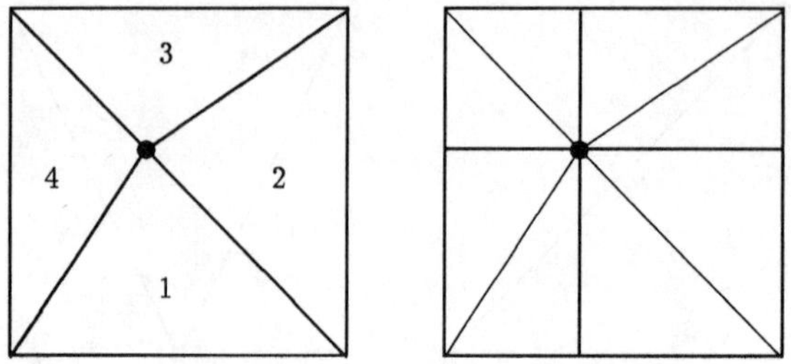

Figure 2: Partitions of Elements for Residual Tests

vertex of each of the four elements as seen in Figure (2). The standard method of Lachat and Watson is applied over the elements (involving two partitions of each element) and the contributions are summed. This method is clearly similar to that described above but involves twice as many subdivisions.

In order to determine which of the methods for singular integration at the internal nodes is most appropriate, integrals are computed for a sequence of residual points on the bisector and on the diagonal of an isoparametric quadratic element as pictured in Figure (3). Integrals for internal nodes are calculated using each of the routines over a single element forming one side of a cube.

Results are given in Tables 1–4 for increasing numbers of Gaussian integration points for both $1/r$ integrals (G-type) and for $1/r^2$ integrals (H-type). Results for the modified Lachat and Watson scheme are listed under the heading 'modified' and those using the conventional routines over each of the sub-elements are headed 'standard'. The columns headed 4×4 *etc.* refer to a 4-point Gauss rule in each dimension over each triangular partition.

As the number of Gauss points is increased the integrals derived from both routines may be seen to converge. The important point to note is that the 'standard' routine results are equivalent to or slightly better than the 'modified' routine results in each case. Indeed, similar results are obtainable by using a 4×4 'standard' rule (8 partitions, 128 Gauss points) to those from a 6×6 'modified' rule (4 partitions, 144 Gauss points). This may have been anticipated as it is generally better to sub-divide an element and apply a lower order rule than to apply an higher order rule.

When isoparametric shape functions are evaluated at mesh nodes the shape function associated with the node is unit-valued, the remainder are zero, so that only one singular integral is generated when computing the $1/r^2$ kernels. This integral need not be calculated explicitly but may be found using the method of row sum elimination as described by Cruse [3]. The one singular integral at the node common to each sub-element in the 'standard' rule described above may be accurately evaluated using row sum elimination. Isoparametric shape functions are in general non-zero when evaluated at a point internal to the element they describe, leading to more than one singular integral, which presumably accounts for the inaccuracies in the H-type integral evaluations for the 'modified' rule.

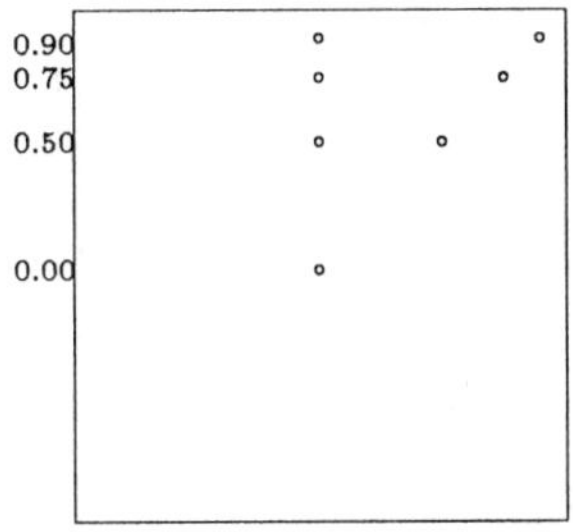

Figure 3: Positions of Nodes for Residual Tests

	4-point		6-point		10-point	
	modified	standard	modified	standard	modified	standard
1	0.6317266	0.6319801	0.6319491	0.6319554	0.6319572	0.6319573
2	0.6196739	0.6216302	0.6218244	0.6220714	0.6220501	0.6220544
3	0.5918487	0.6059384	0.6009976	0.6049494	0.6047435	0.6051769
4	0.5523478	0.5924327	0.5668204	0.5869924	0.5795467	0.5862775

Table 1: G Integral Results as Residual Node Approaches Mid-edge

	4-point		6-point		10-point	
	modified	standard	modified	standard	modified	standard
1	0.6317266	0.6319801	0.6319491	0.6319554	0.6319572	0.6319573
2	0.6037625	0.6021990	0.6025563	0.6027358	0.6027078	0.6027058
3	0.5551298	0.5633214	0.5630217	0.5616302	0.5619403	0.5618802
4	0.5432679	0.5254570	0.5173342	0.5209336	0.5200890	0.5193449

Table 2: G Integral Results as Residual Node Approaches a Vertex

	4-point		6-point		10-point	
	modified	standard	modified	standard	modified	standard
1	0.4409961	0.4414609	0.4404867	0.4404978	0.4405067	0.4405068
2	0.3561520	0.3602880	0.3617677	0.3625624	0.3623117	0.3623272
3	0.2471726	0.2720787	0.2676234	0.2754165	0.2749064	0.2761263
4	0.1436910	0.2029856	0.1555400	0.1812182	0.1695663	0.1810125
5	0.0715954	0.0543338	0.0656455	0.0543485	0.0612073	0.0543464

Table 3: H Integral Results as Residual Node Approaches Mid-edge

	4-point		6-point		10-point	
	modified	standard	modified	standard	modified	standard
1	0.4409961	0.4414609	0.4404867	0.4404978	0.4405067	0.4405068
2	0.4037354	0.4015511	0.4017643	0.4021047	0.4020139	0.4020096
3	0.3145749	0.3237007	0.3333756	0.3313402	0.3310583	0.3309608
4	0.3015195	0.2582300	0.2390548	0.2411229	0.2448376	0.2437120
5	0.1198326	0.1198326	0.1198409	0.1198409	0.1198409	0.1198409

Table 4: H Integral Results as Residual Node Approaches a Vertex

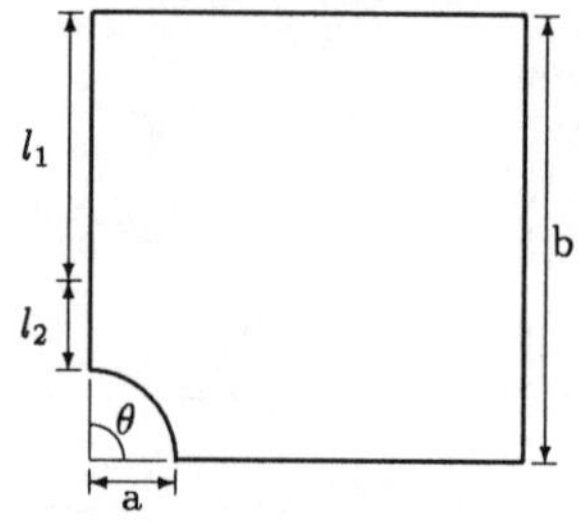

Figure 4: Mesh Dimensions for Test Problem

EXAMPLE PROBLEMS

The Residual Function

The calculation of residuals for isoparametric quadratic elements, using the 'standard' method of integration as described in the previous section, is now considered. A series of residual values across the diagonal of a typical element are computed for two test problems (using a 6 × 6 Gauss point rule) - a cube with a constant tension on opposite faces and a block with a centre hole.

Cube Problem The solution to the cube problem is linear and hence should be represented accurately with quadratic elements. In the centre of an element for which typical results are plotted in Figure (6) the residual is very nearly zero reflecting the small errors in the solution of the problem. However, at the edges, residuals are large. The residuals for higher numbers of Gauss points gave an increased range of small values in the central area and peaks of lower magnitude in each end region.

The peaks for residual values near the edge of the element are believed to be entirely spurious but may have been caused by the rapid changes in the components of the solution here. The constant c_i in Equation (1) is 0.5 for a smooth boundary and 0.25 for an edge, so that the components of the integral equation change rapidly near the edge. Provided that residual points are only chosen in the smooth region away from the element edge it is not necessary to improve the residual estimates in the edge region; particularly since this is likely to be a rather expensive process.

Block Problem No analytical solutions exist for any but the most trivial problems in three-dimensional elastostatics. As a simple example, a three-dimensional approximation to the classical two-dimensional plane strain problem of a hole in an infinite plate under uni-axial tension is considered. The problem was meshed taking account of all symmetries using a mesh similar to that shown in Figure (5). The dimensions of the mesh were based on those defined in Figure (4) with $b/a = 10.0$.

The results for this problem are less encouraging than those for the cube problem because the residual function approximation may be highly oscillatory. It is desirable for the calculation of error estimators and indicators - involving integral norms - to have a smooth approximation over that part of the element from which integration points are chosen. A further problem of residual calculation is where to choose the source points on the element; they must necessarily be few and placement is therefore important, especially if the residual approximation is oscillatory. Residuals calculated close to the edges of an element proved even more unreliable as they did for the cube problem.

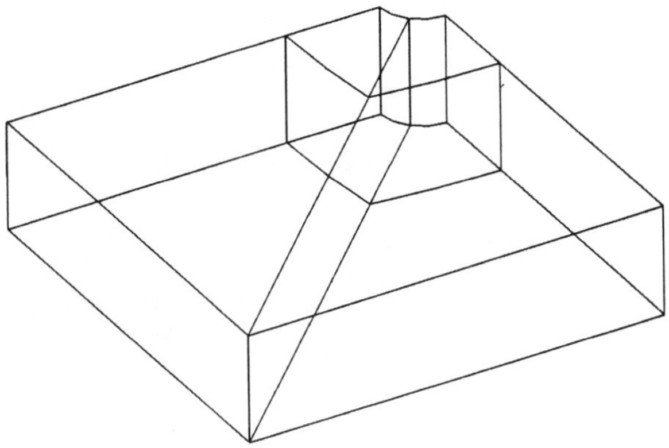

Figure 5: Mesh for block with centre hole under uniaxial tension

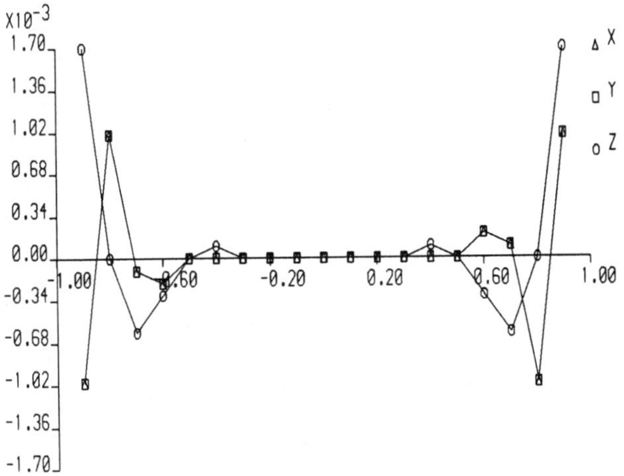

Figure 6: Residuals along diagonal for cube

Mesh	Error Indicator(s)		Stress Concentration
basic	0.8194		1.63
1:1	0.0222	0.3457	2.00
3:1	0.0438	0.1029	2.41
4:1	0.0561	0.0652	2.55
5:1	0.0705	0.0449	2.64
6:1	0.0871	0.0338	2.71

Table 5: Error Indicators and Stress Concentrations

The smoothness of the residual function may be improved slightly by increasing the number of Gauss points and also by improving the evaluation of singular or 'near singular' integrals. 'Near singular' integrals occur when the target element is relatively close to the source point of the integration, so that although r - the distance between target element and source point - is non-zero it may become rather small, causing large changes in terms like $1/r^2$. Errors in evaluating these integrals may be reduced by remeshing to make elements of similar size, which is inefficient, or by using a special integration scheme, such as those given by Telles [8] and Hayami and Brebbia [4]. A simplified method has been adopted, where the target element is sub-divided into four sub-elements with no graduation. This approach gave improvement in the behaviour of the residual function - oscillations are less severe and the curve is smoother.

The most important contributory factor to the smoothness of the residual function however, is the aspect ratio of the element containing the residual node. The sub-division of the element for the numerical integration technique described above creates sub-elements with very much worse aspect ratios than the parent element. Elements with poor aspect ratios have correspondingly poor residual function approximations and those with more reasonable ratios provide smooth approximations. This effect can be seen from Figure (7) where the aspect ratio is 1.1 and Figure (8) with aspect ratio of 2.7. With residual nodes placed as for a 2 × 2 Gauss rule these give, respectively, maximum aspect ratios of 3.7 and 9.1 for the sub-elements.

Error Indicators

Using a series of meshes similar to that shown in Figure (5) error indicators are also calculated for the block with a centre hole problem described above. The block is meshed with two elements along the line of symmetry $\theta = \pi/2$ excepting the 'basic' mesh where only one element is present. The ratio between the elements $l_1 : l_2$ is varied with the smaller element positioned nearer to the hole, where the maximum stress concentration occurs ($\sigma = 3.0$ at $\theta = \pi/2$). Values for error indicators are calculated using Equation (2) with a 2 × 2 rule and the stress concentrations are an average of values taken from the solution. Results are given in Table (5) for residuals computed using a 6 × 6 rule with the error indicator for the element furthest from the hole listed first of all.

The meshes used are simple (16 elements) so that error in values of the maximum stress are large. The correlation of error indicators with the accuracy in the calculated stresses is good and the error indicators would therefore be useful in mesh adaption but when error in the mesh is small they may not be accurate enough to permit further meaningful refinement. As the grading of the mesh is altered the error indicator for the element closest to the hole decreased and that for the element furthest away increased, whilst the value of the maximum stress concentration converged towards the value 3.0. The overall error in the solution increased for the final mesh representing inaccuracy away from the hole and therefore any further optimisation would have to account for this.

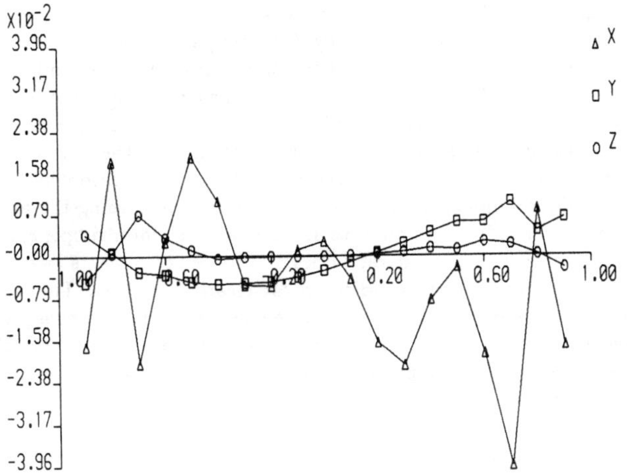

Figure 7: Residuals for block with centre hole (aspect ratio = 2.7)

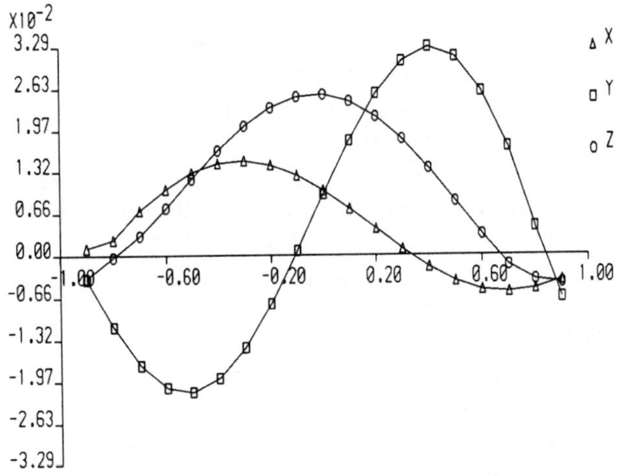

Figure 8: Residuals for block with centre hole (aspect ratio = 1.1)

DISCUSSION

The numerical evaluation of residuals in 3D elasticity is a non-trivial process requiring careful evaluation of both singular and 'near singular' integrals. The aspect ratio of the residual element is also clearly an important factor to consider. It is undesirable, however, to restrict the shape of elements too severely when adapting or optimizing meshes since this would increase the number of elements required. In preference, a more appropriate integration scheme must be adopted. Making further sub-divisions of the residual element, thus reducing the aspect ratios to a more reasonable range, should be relatively simple to implement and more efficient than increasing the number of elements. Using these error indicators to derive an optimal mesh for a simple example gave a considerable improvement in peak stress values, but more experiments to examine correlation between actual errors and error indicators are clearly necessary. The method described here appears to give error indicators that change smoothly and consistently, provided that the aspect ratios are carefully observed, for errors of the order of 5%. It should provide a useful basis for development of adaptive meshing techniques for the solution of 3D elasticity problems.

ACKNOWLEDGEMENT

The authors would like to thank Dr. D. P. Rooke for his interest and encouragement in this project.

REFERENCES

[1] M. H. Aliabadi. An Enhanced Boundary Element Method for Determining Fracture Parameters. In *Proceedings 4th International Conference of Numerical Methods in Fracture Mechanics*, pages 27–39. Pineridge Press, 1987.

[2] T. A. Cruse. Numerical Solutions in Three-dimensional Elastostatics. *International Journal of Solids and Structures*, 5:1259–1274, 1969.

[3] T. A. Cruse. An Improved Boundary Integral Equation Method for Three-dimensional Stress Analysis. *Computers and Structures*, 4:741–754, 1974.

[4] K. Hayami and C. A. Brebbia. A New Coordinate Transformation Method for Singular and Nearly Singular Integrals over General Curved Boundary Elements. In C. A. Brebbia et al, editor, *Boundary Element IX*, pages 375–399, 1987.

[5] J. C. Lachat and J. O. Watson. Effective Numerical Treatment of Boundary Integral Equations: a Formulation for Three-dimensional Elastostatics. *International Journal for Numerical Methods in Engineering*, 10:991–1005, 1976.

[6] P. Parreira. Self-adaptive p-Hierarchical Boundary Elements in Elastostatics. In C. A. Brebbia et al, editor, *Boundary Element IX*, pages 351–373, 1987.

[7] E. Rank. Adaptive Boundary Element Methods. In C. A. Brebbia et al, editor, *Boundary Element IX*, pages 259–278, 1987.

[8] J. C. F. Telles. A Self-adaptive Coordinate Transformation for Efficient Numerical Evaluation of General Boundary Integrals. *International Journal for Numerical Methods in Engineering*, 24:959–973, 1987.

A Posteriori Error Estimates for Linear and Quadratic Boundary Element Formulations

Joseph J. Rencis and Kwo-Yih Jong
Mechanical Engineering Department, Worcester Polytechnic Institute, Worcester, Massachusetts 01609, USA

ABSTRACT

A posteriori error estimates are developed for linear and quadratic boundary element formulations to analyze two-dimensional linear potential problems. Three example problems are considered to demonstrate the accuracy and reliability of the proposed error estimation strategy.

1. INTRODUCTION

Numerical tools such as the finite difference, finite element and boundary element methods are used everyday in industrial environments to solve a wide range of problems. Commercial codes that implement these tools have extensive pre-processing, processing and post-processing capabilities for the user. These three capabilities reduce engineering time dramatically compared to commercial codes ten years ago. Time reduction is associated with the analysis stage of the design process. To compete successfully in a highly competitive marketplace, these codes must yield very accurate and reliable solutions so that high quality products are produced in a short period of time.

Error estimates are very desirable in the everyday engineering environment due to their practicality to save time, and ability to yield accurate and reliable solutions. The error estimates provide a mechanism for an engineer to judge the inaccuracy associated with a numerical tool on the local (element) and global levels. Without error estimates, an engineer must rely on experience obtained through years of modeling practice.

In this paper *a posteriori* error estimates are developed for the boundary element method based on the work of Rencis and Jong [1-5]. Being *a posteriori* in nature, these error estimates are based on the solutions obtained from the boundary element mesh. The error estimates are classified as either an error estimator or error indicator. An error estimator provides a quantitative measure of accuracy associated with the global level, i.e. the entire domain of a problem. On the other hand, an error indicator provides a quantitative measure on a local (element) level. Utilizing the error indicator for each element in the mesh, the error estimator is found by summing the error indicators. The error estimator and error indicator in this work are expressed as a relative percentage, as opposed to actual values, so that one can easily determine the accuracy of a mesh. Since an error indicator is calculated from the numerical solution in the local vicinity of an element, computational effort and storage requirements are small.

In Section 2 an overview of the error estimation strategy employed in this work is discussed. Section 3 highlights various techniques that are employed to determine the accuracy and reliability of the error estimation strategy discussed in Section 2. Three numerical examples are presented in Section 4 and conclusions are stated in Section 5.

2. OVERVIEW OF ERROR ESTIMATION STRATEGY

In this work only problems of the linear potential type governed by the two-dimensional Laplace equation are considered. Due to the boundary element discretization process, one assumes a polynomial variation of the potential u and flux q throughout an element to solve problems of the linear potential type. When a boundary element analysis is carried out for a problem, discretization errors arise for unknown quantities. In other words, if a portion of the surface has a prescribed potential, then the solution found from the boundary element analysis is the flux and this quantity is in error. On the other hand, if the remaining portion of the surface has the flux prescribed, then the boundary element equations solve for the potential and this quantity is in error.

To evaluate errors associated with the unknown potential and flux, one can establish an error estimate. The error estimate provides a mechanism for the engineer to establish the percentage error associated with the potential and flux. The *a posteriori* error estimates developed in this work are based on the boundary element solution as discussed in Rencis and Jong [1-5]. The two types of error estimates are:

- ${}^{0}\eta_u$ - Denotes the predicted relative percentage error of potential u, where the superscript 0 denotes a predicted value. The scalar value ${}^{0}\eta_u$ is expressed as the predicted L_2 boundary error norm of potential ${}^{0}||e_u||$ divided by the predicted potential boundary norm ${}^{0}||u||$.

- ${}^{0}\eta_q$ - Denotes the predicted relative percentage error of flux q. The scalar quantity ${}^{0}\eta_q$ is expressed as the predicted L_2 boundary error norm of flux ${}^{0}||e_q||$ divided by the predicted flux boundary norm ${}^{0}||q||$.

These predicted error estimates can be defined on the element (i.e., called error indicators) and the global (i.e., called error estimator) levels. The error estimator, i.e., predicted global relative percentage error of potential ${}^{0}\eta_u$ is defined according to Rencis and Jong [1-5] as

$$ {}^{0}\eta_u = \left(\frac{{}^{0}||e_u||^2}{||\hat{u}||^2 + {}^{0}||e_u||^2} \right)^{1/2} x\ 100\% \qquad (1a)$$

with the predicted global L_2 boundary error norm for potential ${}^{0}||e_u||$ as

$$ {}^{0}||e_u||^2 = \sum_{i=1}^{n} \left(\int_{\Gamma_i} ({}^{0}e_u)^T \, ({}^{0}e_u) \, d\Gamma \right) = \sum_{i=1}^{n} {}^{0}||e_u||_i^2 \qquad (1b)$$

where summation is over the number of elements, n, on the entire boundary and the superscript T denotes transpose. The boundary element global potential norm $||\hat{u}||$ is defined as

$$ ||\hat{u}||^2 = \sum_{i=1}^{n} \left(\int_{\Gamma_i} \hat{u}^2 \, d\Gamma \right) = \sum_{i=1}^{n} ||\hat{u}||_i^2 \qquad (1c)$$

The predicted error function of potential ${}^{0}e_{u_i}$ for element i, utilized in Equations (1b) is defined as

$$ {}^{0}e_{u_i} = {}^{0}u_i - \hat{u}_i \qquad (1d)$$

where ${}^{0}u_i$ represents a predicted approximation of the potential for element i which is of higher order accuracy than the boundary element potential solution $\hat{u}_i$ for element i.

The known function $\hat{u}_i$ in Equation (1d) is the boundary element solution. However, the predicted potential function ${}^{0}u_i$ in Equation (1d), is not known, so it must be found to determine the predicted potential error function ${}^{0}e_{u_i}$. If an analytical solution is available to solve this problem, then ${}^{0}u$ denotes the exact solution. The function ${}^{0}u_i$ for element i will be evaluated locally based the the boundary element solutions $\hat{u}_i$ of surrounding elements. Consider for simplicity a portion of the boundary as shown in Figure 1 where the flux is prescribed and potential is unknown for elements 1, 2 and 3. Constant elements are shown in Figure 1 for visual clarity in this discussion even though linear and quadratic elements are employed for the error estimation strategy in this work. In order to evaluate the local predicted potential function ${}^{0}u_2$ for element 2, one must look at the potential error of Equation (1d) since the prescribed boundary condition of element 2 is flux.

After carrying out a boundary element analysis, the potential boundary element $\hat{u}$ and exact u solutions can be plotted over the three constant boundary elements shown in Figure 1. To obtain an acceptable solution for the predicted potential function ${}^{0}u$, one can use a local extrapolation process in which ${}^{0}u$ is interpolated by a linear shape function ${}^{0}\mathbf{N}$ of one order higher than the assumed boundary element variations, i.e.

$$ {}^{0}u = {}^{0}\mathbf{N}\, {}^{0}\mathbf{u}' \tag{2} $$

where ${}^{0}\mathbf{u}'$ represents a column vector of predicted mesh point potentials and ${}^{0}\mathbf{N}$ consists of linear shape functions. When a linear or quadratic potential boundary element solution $\hat{u}$ is utilized, then the shape functions that define ${}^{0}u$ are assumed a quadratic or cubic polynomial distribution, respectively.

The problem is to evaluate the predicted mesh point potentials ${}^{0}\mathbf{u}'$ in Equation (2), that are denoted by hollow circles in Figure 1, which best fit, in a least squares sense the boundary element potential solution $\hat{u}$ as demonstrated in Rencis and Jong [1-5]. The solution determined at points B and C (hollow circles) in Figure 1 are used to evaluate the predicted potential function ${}^{0}u_2$ to find the predicted error function in Equation (1d). The predicted percentage error for flux ${}^{0}\eta_q$ is defined in a similar fashion to ${}^{0}\eta_u$ by replacing the subscript u with q in Equations (1a) through (1d).

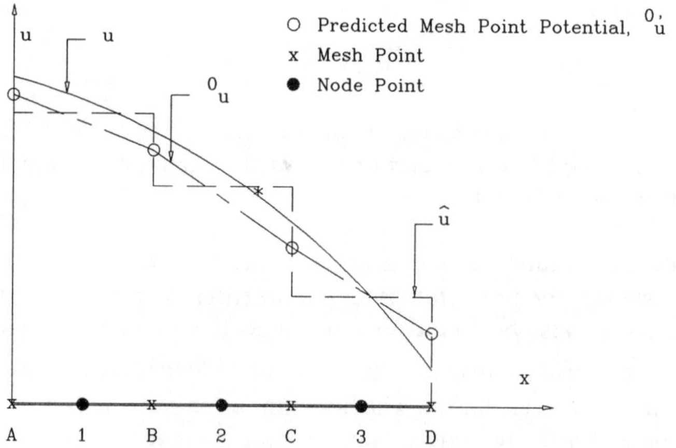

Figure 1. Variations of u, ${}^{0}u$ and $\hat{u}$.

To increase the accuracy of the predicted relative percentage errors of potential ${}^{0}\eta_u$ and flux ${}^{0}\eta_q$, an empirical multiplying factor has been employed for potential F_u and flux F_q, respectively. The corrected values are:

- ${}^{*}\eta_u = F_u\, {}^{0}\eta_u$ - Denotes the corrected relative percentage error of potential u and the superscript * denotes a corrected value. The empirical correction factor F_u depends on the element type used in an analysis and is 0.9 for constant, 0.9 for linear and 0.75 for quadratic element types.

- ${}^{*}\eta_q = F_q\, {}^{0}\eta_q$ - Denotes the corrected relative percentage error of flux q. The empirical correction factor F_q depends on the element type and is 1.0 for constant, 1.8 for linear and 1.65 for quadratic.

The empirical correction factors F_u and F_q were determined by considering various simple and complex problems and it was determined that they are the most appropriate factors to achieve the best accuracy [4,5].

3. RELIABILITY AND ACCURACY MEASURES OVERVIEW

To study the accuracy and reliability of the error estimates discussed in Section 2, exact relative percentage errors and effectivity indices are found. The exact relative percentage error values are found according to Rencis and Jong [1-5] as:

- η_u - Denotes the exact (analytical) relative percentage error of potential. The scalar value η_u is expressed as the exact L_2 boundary error norm of potential $||e_u||$ divided by the exact boundary potential norm $||u||$.

- η_q - Denotes the exact relative percentage error of flux. The scalar quantity η_q is expressed as the exact L_2 boundary error norm of flux $||e_q||$ divided by the exact boundary flux norm $||q||$.

These percentage errors for potential and flux can be found on the local (element) and global levels for the exact η values.

The exact global relative percentage error of potential η_u is defined as [1-5]

$$\eta_u = \frac{||e_u||}{||u||} \tag{3a}$$

where the exact L_2 global boundary potential error norm $||e_u||$ is defined as

$$||e_u||^2 = \sum_{i=1}^{n} \left(\int_{\Gamma_i} (e_u)^T (e_u)\, d\Gamma \right) = \sum_{i=1}^{n} ||e_u||_i^2 \tag{3b}$$

where $e_u = u - \hat{u}$ is the exact potential error function, u is the exact potential function and $\hat{u}$ is the boundary element solution. Furthermore, the exact potential norm $||u||$ is defined as

$$||u||^2 = \int_\Gamma u^2 \, d\Gamma = \sum_{i=1}^{n} \left(\int_{\Gamma_i} (u_i)^2 \, d\Gamma \right) = \sum_{i=1}^{n} ||u||_i^2 \tag{3c}$$

To evaluate the exact expressions in Equations (3a) through (3c), extremely fine boundary boundary element meshes are utilized in accordance to Shephard [6]. The mathematical expression for the exact flux relative percentage error η_q is defined in a similar manner to the exact potential relative percentage error η_u if one replaces the subscripts u with q in Equations (3a) through (3c).

To measure accuracy and reliability of the predicted L_2 boundary potential error norm $^0||e_u||$ defined in Equation (1b), one can define the effectivity indices for potential θ_u and flux θ_q according to Rencis and Jong [1-5] as:

- θ_u - Signifies the effectivity index of the potential. The effective index of potential is expressed as the predicted L_2 boundary potential error norm divided by the exact L_2 boundary potential error norm.

- θ_q - Signifies the effectivity index of the flux. The effective index of flux is expressed as the predicted L_2 boundary flux error norm divided by the exact L_2 boundary flux error norm.

The potential global effectivity index θ_u is defined according to Rencis and Jong [1-5] as

$$\theta_u = \frac{^0||e_u||}{||e_u||} \tag{4}$$

where the predicted L_2 global boundary potential error norm $^0||e_u||$ is defined in Equation (1b) and the exact L_2 global boundary potential error norm $||e_u||$ is defined in Equation (3b).

When the predicted error norm $^0||e||$ and exact error norm $||e||$ are equal, then one can say that the predicted and exact errors are the same, i.e., $\theta = 1$. Noor and Babuška [7] state that an *a posteriori* error estimate is asymptotically correct if the ratio of the error estimate to the true error converges to one as the true error tends to zero (i.e., a necessary requirement is that $\theta \to 1$ as $||e|| \to 0$). This implies that when $\theta \to 1$, either the mesh size $h \to 0$ or polynomial degree $p \to \infty$. This implies that as one refines a mesh by either h or p, then the *a posteriori* error estimate becomes more accurate. The effectivity index θ_q is defined in a similar manner to θ_u if the subscript u is replaced with the subscript q in Equation (4).

4. APPLICATIONS

To verify the accuracy and reliability of the error estimates discussed, computational experiments are carried out for three problems. In this presentation linear and quadratic boundary element formulations are considered.

4.1 Example 1 - Bluff Body

Consider the potential flow around a square body contained between two impermeable boundaries in two-dimensional space as considered by Kelly et al. [8] as shown in Figure 2. For this problem the boundary condition on all sides of the domain is flux. Since the unknown quantity associated with each element is potential, the error due to the boundary element discretization process is potential. To study the accuracy of the error estimates discussed in the Section 2, an initial boundary element mesh consisting of 14 linear elements is shown in Figure 3 and three successively refined meshes are stated.

The first study will be concerned with a linear boundary element mesh as shown in Figure 3 and the second study will be a quadratic boundary element mesh. The quadratic boundary element mesh consists of the same number of elements as the linear boundary element mesh shown in Figure 3. The error estimators for a mesh composed of linear elements are shown in Table 1 and error estimators for a quadratic mesh are shown in Table 2. In Tables 1 and 2, η_u denotes the exact percentage error associated with

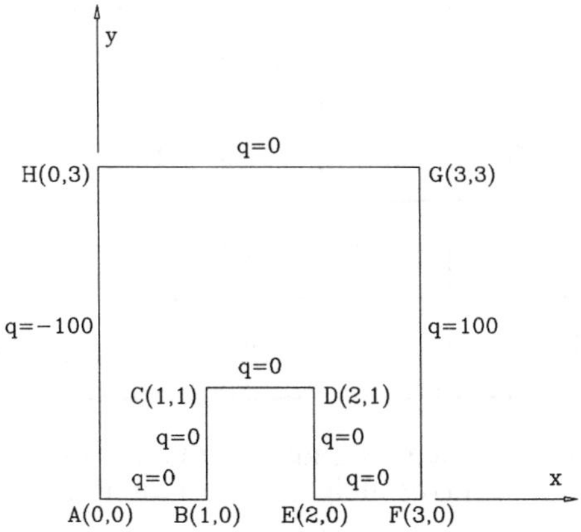

Figure 2. Geometry and boundary conditons for Example 1.

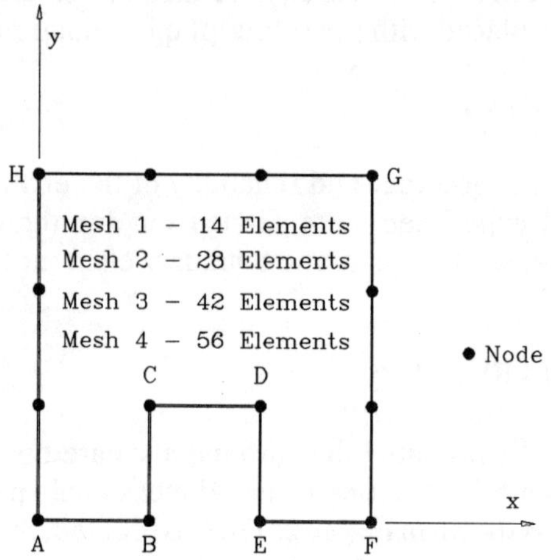

Figure 3. Linear boundary element mesh for Example 1.

potential, ${}^{0}\eta_u$ signifies the predicted percentage error of potential, ${}^{*}\eta_u$ is the corrected percentage error of potential, θ_u is the effectivity index of potential and ${}^{*}\theta_u$ denotes the corrected effectivity index of potential. One should note that the exact η_u is lower for the quadratic mesh as opposed to the linear mesh. This is apparent since the quadratic mesh models this problem more accurately than a linear mesh.

Mesh	# Elem.	η_u	${}^{0}\eta_u$	${}^{*}\eta_u$	θ_u	${}^{*}\theta_u$
1	14	2.9937	3.2530	2.9277	1.1328	1.0195
2	28	1.2396	1.3950	1.2555	1.1419	1.0277
3	42	0.7310	0.8662	0.7796	1.1942	1.0748
4	56	0.4971	0.6188	0.5570	1.2511	1.1261

Table 1. Linear boundary element mesh potential error indicators for Example 1.

Mesh	# Elem.	η_u	${}^0\eta_u$	${}^*\eta_u$	θ_u	${}^*\theta_u$
1	14	1.5195	1.6921	1.2691	1.1262	0.8447
2	28	0.6012	0.7369	0.5527	1.2309	0.9232
3	42	0.3398	0.4588	0.3441	1.3534	1.0151
4	56	0.2183	0.3281	0.2461	1.5047	1.1285

Table 2. Quadratic boundary element mesh potential error indicators for Example 1.

From Tables 1 and 2, one observes that the exact η_u and corrected ${}^*\eta_u$ are in very good agreement for the coarse mesh (Mesh 1) through the fine mesh (Mesh 4). This demonstrates that the percentage error of a mesh consistently decreases as more elements are utilized. Furthermore, the error estimator is also very reliable for the coarse (Mesh 1) through the fine mesh (Mesh 4) as indicated by the θ's. The corrected ${}^*\theta_u$'s in Tables 1 and 2 compare favorably to the exact value of 1.0. One should note that the predicted potential function ${}^0u$ in Equation (1d) was assumed one order higher then the boundary element solution. In other words, for the linear boundary element mesh, the predicted potential function ${}^0u$ was assumed a quadratic function and for the quadratic mesh a cubic function for ${}^0u$ was assumed.

4.2 Example 2 - I Beam

The second example is the heat conduction in a I beam [9] as shown in Figure 4. All surfaces are insulated except the upper and lower faces of the top (side AB) and bottom (side GH) flanges, respectively. The initial two-dimensional linear boundary element mesh employed is shown in Figure 5 and two other successively refined meshes are stated. The quadratic boundary element mesh has the same number of elements as the linear mesh shown in Figure 5. Since this is a mixed boundary value problem, errors will be present for the flux q along sides AB and GH and the potential u has error along the other sides.

In Tables 3 through 6 the global error estimators η and effectivity indices θ for the potential and flux are shown, respectively, for three meshes. Tables 3 and 4 are for the linear boundary element mesh and Tables 5 and 6 are for quadratic boundary element mesh. The accuracy of the corrected error percentage for the potential ${}^*\eta_u$ versus the exact η_u are very good as shown in Tables 3 and 5 for the linear and quadratic meshes, respectively. These results demonstrate the consistency of the error estimator for Mesh 1 through the Mesh 3. The corrected effectivity index $\overset{*}{\theta}_u$ are in the vicinity of the exact value of 1.0. One should note that the predicted potential function ${}^{0}u$ in Equation (1d) was assumed to be quadratic for the linear mesh, and cubic for the quadratic mesh.

The error percentages η_q and effectivity indices θ_q once again are in good agreement for the flux as shown in Tables 4 and 6 for the linear and quadratic meshes, respectively. The predicted flux function ${}^{0}q$ was assumed to be linear for the linear mesh, and quadratic for the quadratic mesh.

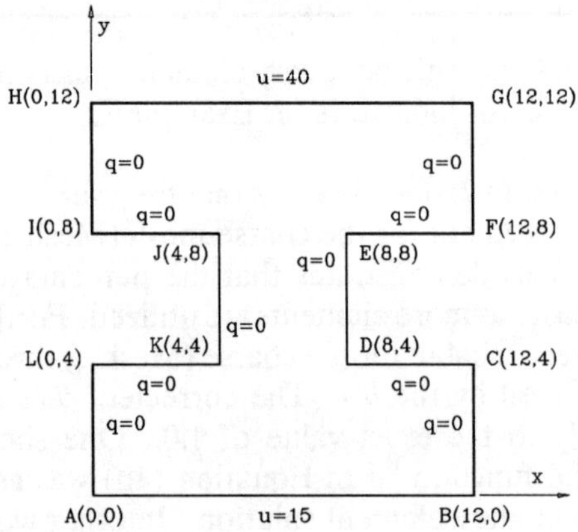

Figure 4. Geometry and boundary conditions for Example 2.

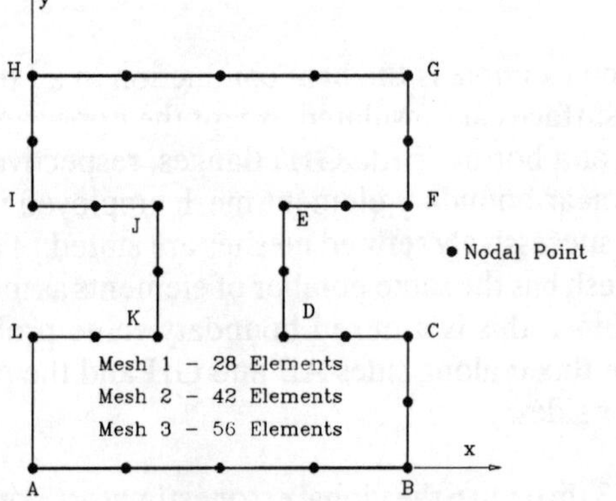

Figure 5. Linear boundary element mesh for Example 2.

Mesh	# Elem.	η_u	${}^0\eta_u$	${}^*\eta_u$	θ_u	${}^*\theta_u$
1	28	0.5518	0.5756	0.5180	1.3385	1.2046
2	42	0.3688	0.3666	0.3299	1.2747	1.1472
3	56	0.2013	0.2210	0.1989	1.1315	1.1018

Table 3. Linear boundary element mesh potential error indicators for Example 2.

Mesh	# Elem.	η_q	${}^0\eta_q$	${}^*\eta_q$	θ_q	${}^*\theta_q$
1	28	3.0944	2.3945	4.3100	0.7823	1.4081
2	42	2.2896	0.9009	1.6216	0.3952	0.7113
3	56	1.1587	0.7510	1.3518	0.6321	1.1379

Table 4. Linear boundary element mesh flux error indicators for Example 2.

Mesh	# Elem.	η_u	${}^0\eta_u$	${}^*\eta_u$	θ_u	${}^*\theta_u$
1	28	0.2670	0.3107	0.2331	1.4919	1.1189
2	42	0.1551	0.1943	0.1457	1.6057	1.2043
3	56	0.1012	0.1391	0.1043	1.7627	1.3219

Table 5. Quadratic boundary element mesh potential error indicators for Example 2.

Mesh	# Elem.	η_q	${}^0\eta_q$	${}^*\eta_q$	θ_q	${}^*\theta_q$
1	28	0.6816	0.5264	0.8949	0.7883	1.3007
2	42	0.3087	0.1902	0.3233	0.6246	1.0306
3	56	0.1825	0.0825	0.1403	0.4569	0.7539

Table 6. Quadratic boundary element mesh flux error indicators for Example 2.

4.3 Example 3 - Transformer Coil

The heat conduction in a transformer coil considered by Rencis and Mullen [10] is shown in Figure 6. Along sides AB, KL and LA the potential is prescribed as zero and along all other sides the flux is prescribed. The initial linear boundary element mesh is shown in Figure 7 and two successive meshes are stated. Due to the boundary element discretization process, errors will exist for the unknown flux along sides AB, KL and LA and for the unknown potential along the other sides. The error indicators for the potential are shown in Tables 7 and 9 for the linear and quadratic element meshes, respectively, and for the flux in Tables 8 and 10 for the linear and quadratic element meshes, respectively. The accuracy of ${}^*\eta_u$ is very good, while ${}^*\eta_q$ is ok.

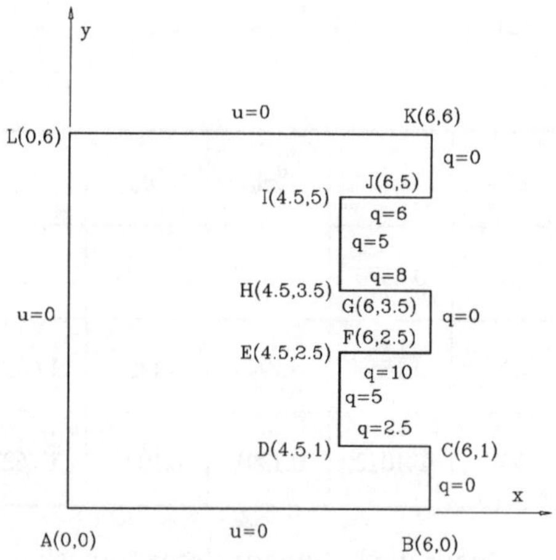

Figure 6. Geometry and boundary conditions for Example 3.

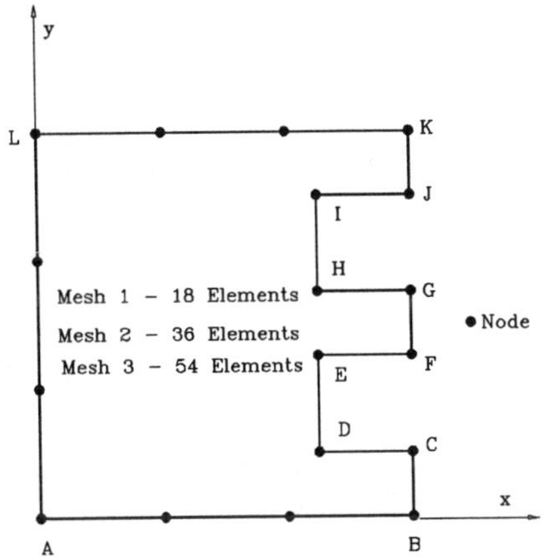

Figure 7. Linear boundary element mesh for Example 3.

One should note that the predicted potential function ${}^{0}u$ in Equation (1d) is quadratic for the linear boundary element mesh, and cubic for the quadratic mesh. In addition, the predicted flux function ${}^{0}q$ is linear for the linear boundary element mesh, and quadratic for the quadratic boundary element mesh.

Mesh	# Elem.	η_u	${}^{0}\eta_u$	${}^{*}\eta_u$	θ_u	${}^{*}\theta_u$
1	18	7.2260	5.5084	4.9576	0.8264	0.7438
2	36	2.6001	2.2387	2.0148	0.8865	0.7978
3	54	1.4681	1.3854	1.2468	0.9609	0.8648

Table 7. Linear boundary element mesh potential error indicators for Example 3.

Mesh	# Elem.	η_q	${}^{0}\eta_q$	${}^{*}\eta_q$	θ_q	${}^{*}\theta_q$
1	18	3.4634	0.9340	1.6812	0.4426	0.7966
2	36	1.5023	0.7010	1.2675	0.7506	1.3511
3	54	0.7937	0.2987	0.5376	0.6035	1.0864

Table 8. Linear boundary element mesh flux error indicators for Example 3.

Mesh	# Elem.	η_u	${}^0\eta_u$	${}^*\eta_u$	θ_u	${}^*\theta_u$
1	18	2.2893	2.5252	1.8939	1.1208	0.8406
2	36	0.8413	1.1598	0.8699	1.3861	1.0355
3	54	0.4055	0.7310	0.5483	1.8073	1.3555

Table 9. Quadratic boundary element mesh potential error indicators for Example 3.

Mesh	# Elem.	η_q	${}^0\eta_q$	${}^*\eta_q$	θ_q	${}^*\theta_q$
1	18	2.0468	0.7407	1.2592	0.5797	0.9565
2	36	0.3794	0.1713	0.2911	0.7223	1.1918
3	54	0.1257	0.0593	0.1008	0.7546	1.2451

Table 10. Quadratic boundary element mesh flux error indicators for Example 3.

5. CONCLUSIONS

An accurate, reliable and computationally inexpensive *a posteriori* error estimation strategy has been developed for the boundary element method. This strategy can be utilized to analyze two-dimensional linear potential problems modeled with linear and quadratic elements.

6. ACKNOWLEDGEMENTS

The authors would like to thank Professor Kenneth E. Scott, Head of the Mechanical Engineering Department at Worcester Polytechnic Institute (WPI) for providing the funds to attend this conference. The authors also appreciate the use of the Computational Mechanics Laboratory in the Mechanical Engineering Department at WPI for helping with this research effort.

7. REFERENCES

1. Rencis, J.J. and Jong, K-Y., An Accuracy Postprocessor for Boundary Element Analysis, in Boundary Elements X (Ed. Brebbia, C.A.), Vol. 1, pp. 187-203, Proceedings of 10th International Conference on the Boundary Elements Method, Southampton, U.K., 1988. Computational Mechanics Publications, Southampton, 1988.

2. Rencis, J.J. and Jong, K-Y. A Self-Adaptive Boundary Element Technique for 2-D Potential Analysis, First Joint Japan/U.S. Symposium on Boundary Element Methods (Ed. Tanaka, M. and Cruse, T.A.), Pergamon Press, Oxford, 1988.

3. Rencis, J.J. and Jong, K-Y. Error Estimates for Elastostatic Boundary Element Solutions, in BETECH 89 (Ed. Brebbia, C.A. and Zamani, N.G.), Proceedings of 4th International Conference on Boundary Element Technology, Windsor, Canada, 1989. Computational Mechanics Publications, Southampton, 1989.

4. Rencis, J.J. and Jong, K-Y. Error Estimation for Boundary Element Analysis, To appear in the ASCE Journal of Engineering Mechanics, September 1989.

5. Rencis, J.J. and Jong, K-Y. A Self-Adaptive h-Refinement Technique for the Boundary Element Method, To appear in Computer Methods in Applied Mechanics and Engineering, 1989.

6. Shephard, M.S. Approaches to the Automatic Generation and Control of Finite Element Meshes, Applied Mechanics Review, Vol. 41, No. 4, pp. 169-185, 1988.

7. Noor, A.K. and Babuška, I. Quality Assessment and Control of Finite Element Solutions, Finite Elements in Analysis and Design, Vol. 3, pp. 1-26, 1987.

8. Kelly, D.W., Mills, R.J., Reizes, J.A. and Miller, A.D. *A Posteriori* Estimates of the Solution Caused by Discretization in the Finite Element, Finite Difference and Boundary Element Methods, International Journal for Numerical Methods in Engineering, Vol. 24, pp. 1921-1939, 1987.

9. Chapman, A.J. Heat Transfer, Macmillan Publishing Company, New York, p. 169, 1984.

10. Rencis, J.J. and Mullen, R.L. A Self Adaptive Mesh Refinement Technique for the Boundary Element Solution of the Laplace Equation, Computational Mechanics, Vol. 3, pp. 309-319, 1988.

REFERENCES

[illegible]

SECTION 3 - NUMERICAL METHODS

An Integration Method for Two-Dimensional Potential Problems Over Curvilinear Geometries

C.V. Camp* and G.S.. Gipson**
*Department of Civil Engineering, Memphis State University, Memphis, TN, USA 38152
**School of Civil Engineering, Oklahoma State University, Stillwater, OK, USA 74078

INTRODUCTION

Rapid, accurate, and consistent evaluations of surface integrals are essential to the successful application of the boundary element method (BEM) to engineering analysis. For a curvilinear surface, a primary source of error is the approximation used to represent the boundary geometry and the variation of the field function. If the geometric boundary is modelled with linear segments, the resulting integrals may be computed exactly. However, the poor approximation of the surface can be a significant source of error in the solution. The use of a more appropriate representation of the boundary, e.g. a quadratic curve, can improve the geometric approximation. Unfortunately, the resulting boundary integrals are much more difficult to evaluate exactly. In this case, the integrations are calculated using some form of Gaussian quadrature.

In this work, a series of exact expressions for several special cases of boundary integrals involving the Laplacian fundamental solution and its normal derivative over a general two-dimensional curvilinear surface are presented. The concept of a quadratic isoparametric representation, usually applied to the variation of the field variables and the geometry, is extended to the parametric mapping of the curvilinear geometry. The result renders the typically complicated Jacobian function into a series of polynomial expressions based on the shape function set and several discrete Jacobian values, Aliabadi[1], Daman, et al[2]. A sequence of exact expressions are developed by substituting the parametric approximation of the Jacobian into the boundary integrations. Implementation of this type of approximation improves the accuracy of the boundary integral solution by eliminating error associated with numerical quadrature. Exact expressions result in consistent accurate evaluations of the field variables at or near the boundary. Overall computational efficiency is improved by reducing the time necessary to compile the system matrices and evaluate the field variables at internal points.

An application of the parametric approximation of the Jacobian using a quadratic element representation is developed. This method can similarly be applied to other boundary element representations, e.g. cubic or Overhauser. A numerical solution of the boundary integral equation for phenomena governed by the Poisson equation is presented and compared with an exact analysis.

FORMULATION

Consider the general form of the Poisson equation in a two-dimensional domain Ω bounded by a surface Γ

$$\nabla^2\phi = f(x,y) \tag{1}$$

where the function f(x,y) is a known function of the spatial coordinates. The boundary integral equation representation of Equation (1) is well known (Brebbia[3], Gipson[4]). For a general field point, the integral equation is

$$ß(p)\phi(p) = \int_\Gamma [\ \phi(q)G'(q,p) - \phi'(q)G(q,p)\ ]\ d\Gamma + \int_\Omega f(x,y)G(q,p)\ d\Omega \tag{2}$$

The vector **p** may locate a point on the surface Γ or in the domain Ω, the vector **q** locates a point on the surface Γ only and may coincide with **p**, the function G(**q**,**p**) is the Laplacian fundamental solution, ϕ(**q**) is the potential function, and the primes denotes derivatives with respect to the outward surface normal. The value of the parameter ß(**p**) is 1.0 for a point inside the domain, some fractional value on the boundary, and zero outside the domain (Brebbia[3]). The fundamental solution and its normal derivative have the form

$$G(q,p) = \frac{1}{2\pi}\ln|q-p| \tag{3}$$

$$G'(q,p) = \frac{(x-x_p)n_x + (y-y_p)n_y}{2\pi|q-p|^2} \tag{4}$$

The variation of the potential function ϕ, its normal derivative, and Cartesian coordinates of vectors **p** and **q** can be approximated over a segment of the boundary by using the following approximations

$$\phi = \langle N\rangle\{\phi\} \qquad \phi' = \langle N\rangle\{\phi'\} \tag{5a,b}$$

$$x = \langle N\rangle\{x\} \qquad y = \langle N\rangle\{y\} \tag{6a,b}$$

where {ϕ}, {ϕ'}, {x}, and {y} are column vectors containing the discrete values of each of the quantities at nodal points on an element. The row vector, <N>, is an assumed distribution for the quantities over each boundary element. A discrete integral expression for the governing equation is formed by substituting Equations (5)-(6) into Equation (2).

$$\sum_{j=1}^{n}\left(\int_{\Gamma_j} [\ \langle N\rangle G'(q,p)\{\phi\} - \langle N\rangle G(q,p)\{\phi'\}\ ]d\Gamma_j + \int_\Omega f(x,y)\ G(q,p)\ d\Omega\right) - ß(p)\phi(p) \tag{7}$$

The summation in Equation (7) is over n elements defining the boundary Γ. The resulting set of equations obtained from Equation (7) require the evaluation of integrals of the form

$$G_{pe} = \frac{1}{2\pi} \int_{\Gamma_e} \langle N \rangle \ln|q-p| \, d\Gamma_e \tag{8}$$

$$H_{pe} = \frac{1}{2\pi} \int_{\Gamma_e} \langle N \rangle \frac{(x-x_p)n_x + (y-y_p)n_y}{2\pi|q-p|^2} \, d\Gamma_e \tag{9}$$

$$B_p = \frac{1}{2\pi} \int_{\Omega} f(x,y) \ln|q-p| \, d\Omega \tag{10}$$

If a linear shape function set is used to describe the surface geometry and the variation of the potential at the surface, the integrals defined in Equations (8) and (9) may be evaluated exactly (Kelmanson[5], Camp and Gipson[6]). For more complex geometric representations, the boundary integrals are very difficult to evaluate analytically. The matter is complicated further when points denoted by vectors **q** and **p** denote points on the same boundary element. In this case, the vector $|\mathbf{q}-\mathbf{p}|$ will be zero at a point along the boundary element and the integrals in Equations (8) and (9) will be singular.

QUADRATIC ELEMENTS

Consider a general quadratic curve defined by a set of three consecutive discrete points, labeled "i", "j", and "k", as shown in Figure 1. The curve description may be transformed from two-dimensional coordinates to a single curvilinear parameter, t. The parametric form for the quadratic shape functions over an interval [0,1] are shown graphically in Figure 2 and given as

$$N_i = 2t^2-3t+1 \qquad N_j = -4t^2+4t \qquad N_k = 2t^2-t \tag{11a,b,c}$$

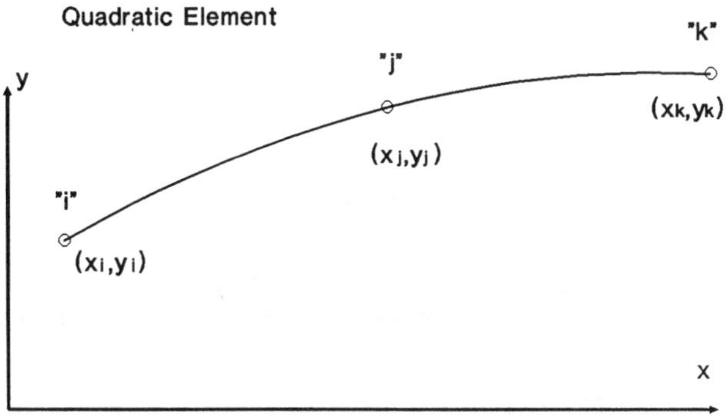

Figure 1. Quadratic element notation

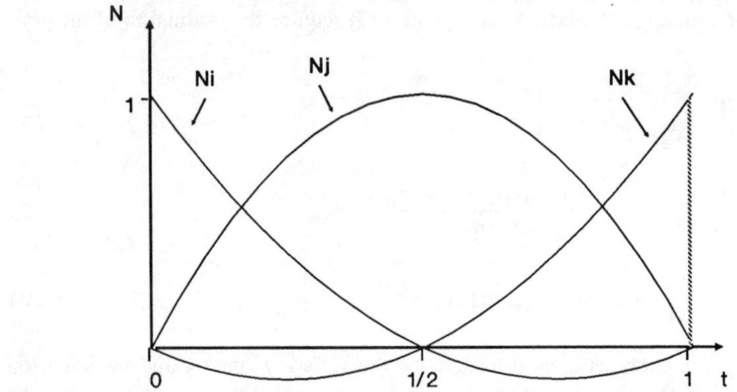

Figure 2. Quadratic element shape functions

An isoparametric boundary element representation may be defined by substituting the shape function set, given in Equations (11a,b,c), into Equations (5) and (6). The resulting distribution of the Cartesian coordinates x and y as a function of the parameter t are

$$x(t) = x_i N_i + x_j N_j + x_k N_k \quad (12a)$$

$$y(t) = y_i N_i + y_j N_j + y_k N_k \quad (12b)$$

Equations (12a) and (12b) may be rewritten as

$$x(t) = A_x t^2 + B_x t + x_i \quad (13a)$$

$$y(t) = A_y t^2 + B_y t + y_i \quad (13b)$$

where the element constants A_x, A_y, B_x, B_y are defined in the Appendix. The integrals defined in Equations (8) and (9) are transformed into the parameter space, t, using the simple Jacobian given as

$$|J| = \left[\left(\frac{dx}{dt} \right)^2 + \left(\frac{dy}{dt} \right)^2 \right]^{1/2} = \frac{d\Gamma}{dt} \qquad d\Gamma = |J|dt \quad (14)$$

A parametric form of the Jacobian may be obtained by calculating the derivatives of x and y, given in Equations (13a,b), with respect to t and substituting them into Equation (14)

$$|J| = \left[4At^2 + 2Bt + C \right]^{1/2} \quad (15)$$

where the constants A, B, and C are given in the Appendix. The position vector $|\mathbf{q}-\mathbf{p}|^2$ may be written in a parametric form by using the relationships defined in Equation (13) as

$$|q-p|^2 \equiv X = At^4 + Bt^3 + (C + E)t^2 + Ft + D \tag{16}$$

where the element constants D, E, and F are given in the Appendix. Substituting the parametric expressions for the Jacobian, $|J|$, the position vector, $|q-p|$, and quadratic shape functions into Equations (8) and (9) results in the following expressions

$$G_{pe} = \frac{1}{4\pi} \int_0^1 \langle N \rangle \ln X \; |J| \; dt \tag{17}$$

$$H_{pe} = \frac{1}{2\pi} \int_0^1 \langle N \rangle \frac{(A_x t^2 + B_x t + D_x) n_x + (A_y t^2 + B_y t + D_y) n_y}{X} |J| \; dt \tag{18}$$

The direction cosines n_x and n_y are functions of the parameter t, the Jacobian $|J|$, and the element constants listed in the Appendix.

$$n_x = \frac{2A_y t + B_y}{|J|} \qquad n_y = -\frac{2A_x t + B_x}{|J|} \tag{19a,b}$$

The parametric form of the boundary integrals, given in Equations (17) and (18), are very difficult to evaluate analytically due to the complex nature of the Jacobian and the fundamental solution. Generally, these integrals are evaluated numerically with Gaussian quadrature. However, if the Jacobian can be accurately approximated as a polynomial, the integrals may be evaluated using exact expressions.

JACOBIAN APPROXIMATION

The transformation from two-dimensional Cartesian coordinates to the parametric space t is accomplished by the simple Jacobian given in Equation (15). This expression may be represented by a quadratic curve of the form presented in Figure 1. The parametric form of the Jacobian is represented by the quadratic shape functions and three consecutive discrete values of the Jacobian. The approximation of the Jacobian, $|J|$, has the form

$$|J| = N_i |J|_{t=0} + N_j |J|_{t=\frac{1}{2}} + N_k |J|_{t=1} \tag{20}$$

where

$$|J|_{t=0} = C^{\frac{1}{2}} \qquad |J|_{t=\frac{1}{2}} = (A + B + C)^{\frac{1}{2}} \tag{21a,b}$$

$$|J|_{t=1} = (4A + 2B + C)^{\frac{1}{2}} \tag{21c}$$

The constants A, B, and C are quadratic element constants and are listed in the Appendix. The quadratic approximation of the Jacobian, Equation (20), is graphically compared to the parametric form of the Jacobian, given by Equation (15), for two cases. The first case, shown in Figure 3(a), compares the Jacobian approximation to the exact form for the relatively "flat" boundary element shown in Figure 1. In Figure 3(b), the Jacobian approximation for a quadratic boundary element representing a quarter circle of unit radius is shown. The results indicate that the quadratic Jacobian representation defined in Equation (20) is capable of accurately approximating the Jacobian for a variety of geometric shapes.

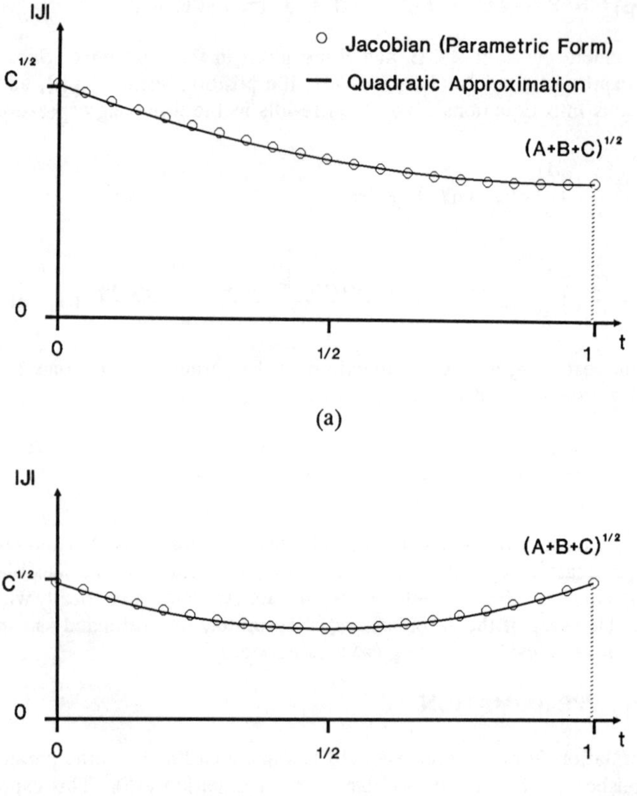

Figure 3. Quadratic approximation of the Jacobian transformation: (a) for an element with small curvature; (b) an element with high curvature

The Jacobian component, |J|, of the integral expressions, given in Equations (17) and (18), may be replaced with the quadratic approximation of Equation (20) and rewritten as

$$G_{pe} = \frac{1}{4\pi}\int_0^1 \langle N\rangle \ln X \langle N\rangle\{J\}\, dt \tag{22}$$

$$H_{pe} = \frac{1}{2\pi}\int_0^1 \langle N\rangle \frac{(A_x t^2+B_x t+D_x)n_x + (A_y t^2+B_y t+D_y)n_y}{X} \langle N\rangle\{J\}dt \tag{23}$$

The <N>{J} term of Equations (22) and (23) may be expanded by multiplying the quadratic shape functions, <N>, and the discrete Jacobian values, {J}, together. For example, Equation (22) may be written as

$$G_{pe} = \frac{1}{4\pi}\int_0^1 \Big[(N_iN_iJ_i + N_iN_jJ_j + N_iN_kJ_k)\ \ln X + (N_jN_iJ_i + N_jN_jJ_j + N_jN_kJ_k)\ \ln X + (N_kN_iJ_i + N_kN_jJ_j + N_kN_kJ_k)\ \ln X \Big] dt \qquad (24)$$

The Jacobian, $|J|$, component present in both the expressions for the direction cosines given in Equations (19a,b) and the integral defined in Equation (24) will cancel, resulting in the following expression

$$H_{pe} = \frac{1}{2\pi}\int_0^1 < N_i \quad N_j \quad N_k > \frac{Y(t)}{X}\, dt \qquad (25)$$

where

$$Y(t) = (A_xt^2+B_xt+D_x)(2A_yt+B_y) - (A_yt^2+B_yt+D_y)(2A_xt+B_x) \qquad (26)$$

NON-SINGULAR CASE

For surface integrations about the general point, (x_p,y_p), which does not coincide with any of the nodes defining the boundary segment under consideration, the integrals in Equations (24) and (25) are not singular. For this case, the integrals require evaluation of terms of the form t^n/X and $t^n\ln X$. Recall, the function X is a quartic equation as defined in Equation (16). The quartic equation, X(t), may be factored into a pair of quadratic equations $X_1 = A_1t^2 + B_1t + C_1$ and $X_2 = A_2t^2 + B_2t + C_2$. Using the bi-quadratic form, Equation (24) contains integrands of the form $t^nX_{1,2}$ which may be evaluated using the following expressions (Camp and Gipson[7])

$$L_0 = \int_0^1 \ln X_{1,2}dt = \ln(A_{1,2}+B_{1,2}+C_{1,2}) - 2A_{1,2}I2 - B_{1,2}I_1 \qquad (27)$$

$$L_n = \int_0^1 t^n\ln X_{1,2}dt = \frac{1}{n+1}\,[\ln(A_{1,2}+B_{1,2}+C_{1,2}) - 2A_{1,2}I_{n+2} - B_{1,2}I_{n+1}] \qquad (28)$$

where $X_{1,2}$ represents X_1 or X_2. The terms I_n are the value of integrals containing terms of the form $t^n/X_{1,2}$ evaluated using the following relationships (Camp and Gipson[7])

$$I_0 = \int_0^1 \frac{dt}{X_{1,2}} = \frac{1}{Q}\ln\frac{2A_{1,2}t + B_{1,2} - Q}{2A_{1,2}t + B_{1,2} + Q} \qquad Q > 0 \qquad (29a)$$

$$= \frac{1}{Q_1}\sin^{-1}\frac{(2A_{1,2}t + B_{1,2})Q_1}{2A_{1,2}(A_{1,2}t^2 + B_{1,2}t + C_{1,2})} \qquad Q < 0 \qquad (29b)$$

$$= \frac{2}{B_{1,2}(1 + B_{1,2}/2C_{1,2})} \qquad Q = 0 \qquad (29c)$$

$$I_1 = \int_0^1 \frac{t}{X_{1,2}} dt = [\ \ln(A_{1,2}+B_{1,2}+C_{1,2}) - \ln C_{1,2} - B_{1,2}I_0]/2A_{1,2} \quad (30)$$

$$I_n = \int_0^1 \frac{t^n}{X_{1,2}} dt = (\frac{1}{n-1} - B_{1,2}I_{n-1} - C_{1,2}I_{n-2})/A_{1,2} \quad (31)$$

where $Q = (B_{1,2}^2 - 4A_{1,2}C_{1,2})^{½}$ and $Q_1 = (4A_{1,2}C_{1,2} - B_{1,2}^2)^{½}$.

Equation (25) contains terms of the form t^n/X where the X(t) is a quartic function. To evaluate t^n/X for a general point (x_p, y_p), the quotient 1/X may be separated using partial factors into a sum of two factions of the form $O(t)/X_1$ and $M(t)/X_2$. Each of the two terms formed by partial factions may be evaluated exactly. However, the error introduced by numerical techniques for determining O(t) and M(t) reduces the accuracy of the exact expressions and may not provide a significant improvement over results using Gaussian quadrature.

SINGULAR CASES

When the point (x_p,y_p), defined by vector **p**, coincides with nodes on the boundary segment, denoted by vector **q**, the integrals in Equations (24) and (25) are singular. For a quadratic element representation of the boundary, there are two cases to consider: first, when (x_p,y_p) is equal to nodes "i" or "k" shown in Figure 1; second, if (x_p,y_p) corresponds to the middle node "j".

In the first case, when $(x_p,y_p) = (x_i,y_i)$, the relationship defined in Equations (16) and (26) reduces to

$$X = At^4 + Bt + Ct^2 \quad (32)$$

$$Y(t) = (A_yB_x - A_xB_y)t^2 \quad (33)$$

The integral defined in Equation (25) may be reduced by substituting the relationships given in Equations (32) and (33) resulting in

$$H_{pe} = \frac{1}{2\pi}\int_0^1 < N_i \quad N_j \quad N_k > \frac{A_yB_x - A_xB_y}{X} dt \quad (34)$$

The reduced form of the position vector X, given in Equation (32), allows the integral in Equation (24) to be rewritten as

$$G_{pe} = \frac{1}{4\pi}\int_0^1 \{N\}<N>\{J\}[2\ln t + \ln(At^2 + Bt + C)]\ dt \quad (35)$$

The singular case integrals defined in Equations (34) and (35) contain terms of the form t^n/X, $t^n\ln t$, and $t^n\ln X$. Equation (34) may be evaluated exactly using the expression defined in Equations (29) - (31). The second term in Equation (35) may be evaluated using Equations (27) and (28). The $t^n\ln t$ term is evaluated using

$$\int_0^1 t^n \ln t \, dt = -1/(n+1)^2 \tag{36}$$

If $(x_p,y_p)=(x_k,y_k)$, the required integrals are identical to those of Equations (34) and (35); however, due to a change of variable, the quadratic shape functions are transformed by the following relationships

$$N_i' = N_k \qquad N_j' = N_j \qquad N_k' = N_i \tag{37}$$

For the second case, when $(x_p,y_p) = (x_j,y_j)$, the integrals in Equations (34) and (35) are rewritten as a function of a new parameter s, that varies from -1 to 1. This effectively divides the integrations in half about the singular point (x_j,y_j). The transformation from t space to s space is given by s = 2t-1. The new quadratic shape functions, position vector, and the Jacobian in parameter s are

$$N_i^* = (s^2-s)/2 \qquad N_j^* = 1-s^2 \qquad N_k^* = (s^2+s)/2 \tag{38a,b,c}$$

$$X^* = A^*s^4 + B^*s^3 + C^*s^2 \tag{39}$$

$$|J|^* = [As^2 + (2A+B)s + A + B + C]^{½} \tag{40}$$

where the transformed element constants A^*, B^*, and C^* are listed in the Appendix. Equation (34) and (35), as a function of the parameter s, are

$$H_{pe} = \frac{1}{4\pi}\int_0^1 < N_i^* \quad N_j^* \quad N_k^* > \frac{A_y^*B_x^* - A_x^*B_y^*}{X^*} \, ds$$

$$- \frac{1}{4\pi}\int_0^{-1} < N_i^* \quad N_j^* \quad N_k^* > \frac{A_y^*B_x^* - A_x^*B_y^*}{X^*} \, ds \tag{41}$$

$$G_{pe} = \frac{1}{8\pi}\int_0^1 \{N^*\}<N^*>\{J^*\}[2\ln t + \ln(A^*t^2 + B^*t + C^*)] \, ds$$

$$- \frac{1}{8\pi}\int_0^{-1} \{N^*\}<N^*>\{J^*\}[2\ln t + \ln(A^*t^2 + B^*t + C^*)] \, ds \tag{42}$$

where the s parameter element constants, denoted by the single asterisk, *, are listed in the Appendix. The second part of each integral in Equations (41) and (42) is transformed into a new parameter space, u, by u = -s. The shape functions, position vector, and the Jacobian over the second half of the element are

$$N_i^{**} = (u^2+u)/2 \qquad N_j^{**} = 1-u^2 \qquad N_k^{**} = (u^2-u)/2 \tag{43a,b,c}$$

$$X^{**} = A^*u^4 - B^*u^3 + C^*u^2 \tag{44}$$

$$|J|^{**} = [Au^2 - (2A+B)u + A + B + C]^{½} \tag{45}$$

Equations (41) and (42) may be rewritten as

$$H_{pe} = \frac{1}{4\pi}\int_0^1 < N_i^* \quad N_j^* \quad N_k^* > \frac{A_y^* B_x^* - A_x^* B_y^*}{X^*} \, ds$$

$$+ \frac{1}{4\pi}\int_0^1 < N_i^{**} \quad N_j^{**} \quad N_k^{**} > \frac{A_y^{**} B_x^{**} - A_x^{**} B_y^{**}}{X^{**}} \, du \qquad (46)$$

$$G_{pe} = \frac{1}{8\pi}\int_0^1 \{N^*\}<N^*>\{J^*\}[2\ln s + \ln(A^* s^2 + B^* s + C^*)] \, ds$$

$$+ \frac{1}{8\pi}\int_0^1 \{N^{**}\}<N^{**}>\{J^{**}\}[2\ln u + \ln(A^* u^2 - B^* u + C^*)] \, du \qquad (47)$$

The element constant indicated by double asterisks, **, is listed in the Appendix. The terms in integrals defined by Equations (46) and (47) may be evaluated using the exact expressions given in Equations (27) - (31).

EXAMPLE ANALYSIS

Consider the torsion of a circular elastic shaft with a radial crack extending from the center to the surface, as shown in Figure 4(a). The problem is governed by the Poisson equation, given in Equation (1), with f(x,y) = -2 and with the potential function, ϕ, representing the torsion function. The surface of the cylinder and the crack are free of shearing stresses; therefore, the value of the torsion function at those surfaces is an arbitrary constant and assumed to be zero. A boundary element mesh composed of 10 quadratic elements, shown in Figure 4(b), which takes advantage of the line of symmetry colinear with the crack in the shaft, is used to model the problem.

The required domain integral representing the loading function f(x,y), given in Equation (10), is transformed into an equivalent series of surface integrals using Green's second identity

$$\int_\Omega f(x,y) G \, d\Omega = \int_\Gamma (f G_1' - f' G_1) \, d\Gamma \qquad (48)$$

where the function G_1 is

$$G_1(q,p) = \frac{1}{8\pi} \ln|q-p|^2 (\ln|q-p| - 1) \qquad (49)$$

The absolute percent error between an exact series solution given by Lebedev, et al.[8] and the results using either an 8-point Gaussian quadrature or the exact expressions is shown in Figure 5. The exact expressions provide a significant improvement in the accuracy of the BEM formulation using 10 quadratic elements. This level of surface discretization, while inadequate for internal points close to the discontinuity in the boundary conditions at the center of the cylinder, provides a good base for comparison of the exact expressions to quadrature. As the number of elements increases, the accuracy of both formulations greatly improves. In this case, the exact expressions provide only a marginal decrease in computational time.

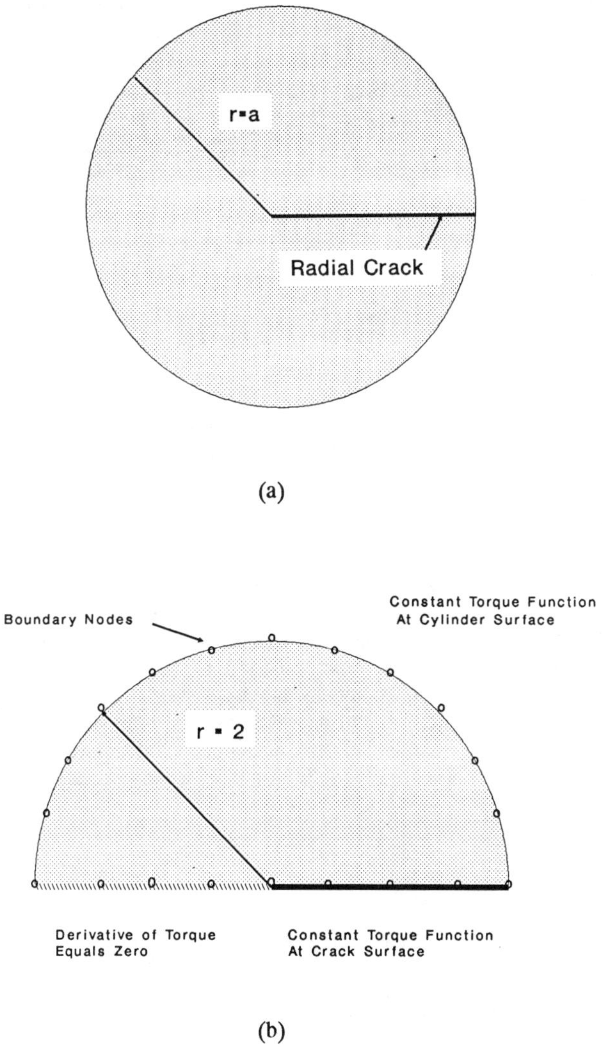

Figure 4. Cylindrical shaft containing a radial crack under constant unit torque: (a) problem definition; (b) boundary element model

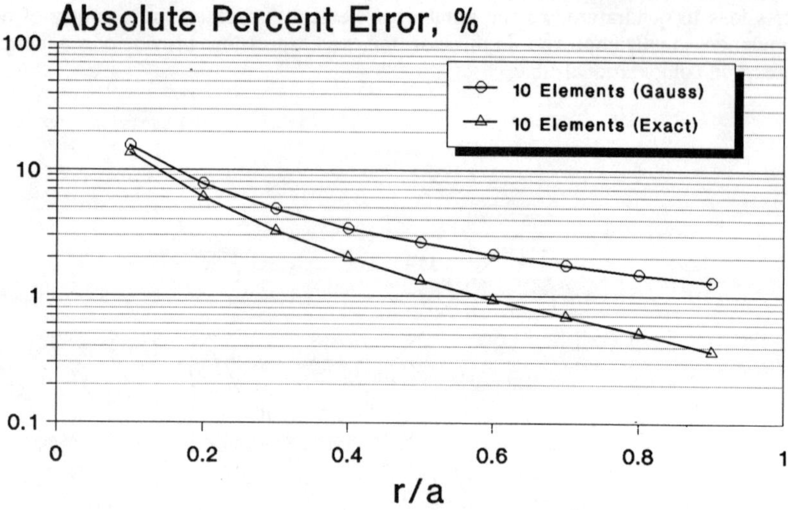

Figure 5. Comparison of values of the torque function along a vertical radius for solutions using Gaussian quadrature and singular exact expressions

SUMMARY

A series of exact expressions for integrals required in a boundary element solution of the Poisson equation are developed. The Jacobian transformation is approximated by a set of quadratic shape functions analogous to that representing the geometry and the variation of the field variable. The exact expressions for the non-singular form of the integrals require several computationally intensive steps. The technique is somewhat slower than Gaussian quadrature and provides minimal improvement in accuracy. However, exact analysis for the singular cases are significantly more accurate than results obtained using Gaussian quadrature while providing a modest reduction in computation time.

APPENDIX

$$A = A_x^2 + A_y^2 \quad B = 2(A_xB_x + A_yB_y) \quad C = B_x^2 + B_y^2$$

$$D = D_x^2 + D_y^2 \quad E = 2(A_xD_x + A_yD_y) \quad F = 2(B_xD_x + B_yD_y)$$

$$A_x = 2x_i - 4x_j + 2x_k \qquad B_x = -3x_i + 4x_j - x_k$$

$$A_y = 2y_i - 4y_j + 2y_k \qquad B_y = -3y_i + 4y_j - y_k$$

$$D_x = x_i - x_p \qquad D_y = y_i - y_p$$

$$A_x^* = (x_i - 2x_j + x_k)/2 \qquad B_x^* = (-x_i + x_k)/2$$

$$A_y^{*} = (y_i - 2y_j + y_k)/2 \qquad B_y^{*} = (-y_i + y_k)/2$$

$$A_x^{**} = A_x^{*} \qquad B_x^{**} = -B_x^{*}$$

$$A_y^{**} = A_y^{*} \qquad B_y^{**} = -B_y^{*}$$

$$A^{*} = \frac{A}{16} \qquad B^{*} = \frac{A}{4} + \frac{B}{8} \qquad C^{*} = \frac{3(A+B)}{8} + \frac{C+E}{4}$$

REFERENCES

1. Aliabadi, M.H. Evaluation of the Integrals in Two-Dimensional Boundary Element Method, Univ. of Southampton Report No. EMR/10/1, 1985.

2. Daman, A.E., Smith R.N.L., and Ellis, L. On the Integration of Isoparametric Quadratic Elements for Curved Boundaries, BETECH 86 (Ed. Connor, J.J., and Brebbia, C.A.), pp. 821-828, Proc. of the 2nd Boundary Element Technology Conference, MIT, USA 1986. Computational Mechanics Publications, 1986.

3. Brebbia, C.A. (Ed.). The Boundary Element Method for Engineers, Pentech Press, London, 1978.

4. Gipson, G.S. Boundary Element Fundamentals - Basic Concepts and Recent Developments in the Poisson Equation, Computational Mechanics, Southampton and Boston, 1987.

5. Kelmanson, M.A. An Integral Equation Method for the Solution of Singular Slow Flow Problems, Journal of Computational Physics, Vol. 51, pp. 139-158, 1983.

6. Camp C.V. and Gipson G.S. A Boundary Element Method for Viscous Flows at Low Reynolds Number, accepted for publication to Engineering Analysis, 1988.

7. Camp, C.V. and Gipson, G.S. Boundary Element Analysis of Nonhomogeneous Biharmonic Phenomena, accepted for publication by Computational Mechanics, Southampton and Boston, 1988.

8. Lebedev, N.N., Skalskaya, I.P., and Ulflyans, Y.S. Worked Problems in Applied Mathematics, translated by R.A. Silverman, Dover Publications, New York, 1965.

Analytical Integration of the Fundamental Solution 1/R Over Panel Boundary Element

R. Rangogni
ENEL-CRIS, 20159 Milano, Italy

ABSTRACT

It is well known that the fundamental solution of the Laplace equation in three dimensional domains is the function 1/R where R denotes the distance from the source point to the integration one.
Using the Boundary Element Method (BEM) the surface of the domain is usually subdivided into a certain number of curved surface elements. The integration of the fundamental solution, and of its normal derivative, usually takes the major part of the CPU time.
Besides more accuracy is required in the integration, when the distance R becomes small (i.e. when the source and integration points lie on the same panel). Many papers have appeared (i.e Cristecu [1], Aliabadi [2], Hayami [3], ecc.) to improve the numerical Gaussian quadrature formulas. In the present paper it is shown that simple analytical integration of function 1/R is possible on plane quadrilateral or triangular panels.
When curved panels are involved, these are approximated by plane subelements and the analytical integration is then performed on each subelements. The analytical integration is not run-time consuming and it will improve the accuracy of the results.

INTRODUCTION

In solving three-dimensional potential problems using Boundary Element Method (BEM) it is important to evaluate accurately the integrals of the fundamental solution 1/R, where R is the distance between source and integration points, because these integrals determine the influence matrices. Whenever distance R is sufficently large the Gaussian quadrature rule works efficiently. However when source and integration points are close or on the same boundary element, Gaussian formulae are not accurate enough and different techniques have to be used.

In this paper we concentrate our attention to the case where source and integration points lie on the same boundary element. The element can be plane or curved, such as isoparametric element used to model curved surfaces. The integration on this element is important and a lot of papers have been written on the subject. A very good paper, with many references was written by Hayami [3] which we refer to.

For planar element an analytical integration of 1/R is possible while curved elements are approximated by planar subelements where the analytical integration is performed.

ANALYTICAL INTEGRATION

We consider a general curved boundary surface element in three dimensions such as the one shown in fig.1 and the case where the source point P and the integration point Q are on the same element and internal to it.

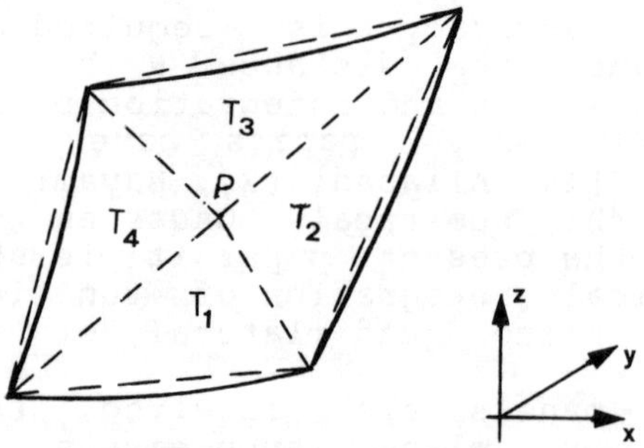

Fig.1 General curved boundary surface element and its approximation by 4 plane triangles.

The curvature of the element should not be very large in such a way that we can approximate the element by means of 4 plane triangles as shown in fig.1. Another way of simplifying the element could be to project the element on some tangent plane as done in Hayami [3]. We can also use more than 4 triangles if better accuracy is required. We deal with the case where the singular point is internal

to the element but treatment for the case when singular point is in a different position or on the boundary of the element is staightforward.
The boundary element formulation for three dimensional potential problems imply to compute the following integral:

$$g = \int_{S_e} \frac{ds}{R} \tag{1}$$

Where S_e is a boundary element. Some analytical formulae have been previously presented but a different approach is here used. Integral (1) is encountered, for example, in solving the Laplace problem in 3-D when one is studying potential problems in water waves (eg. Newman [4]) or in aerodynamics (eg. Hess [5]). As said above we divide the planar boundary element into 4 subelements as shown in fig.1. Integral (1) then becomes

$$g = \int_{S_e} \frac{ds}{R} = \sum_{k=1}^{4} \int_{T_k} \frac{dT}{R} = \sum_{k=1}^{4} I_4 \tag{2}$$

where T_k, k=1,2,3,4 denotes each triangle in fig.1. It is an easy task to construct a local topology of each triangle starting from the topology of the quadrilateral boundary element so we do not go into details. If the boundary element is already a triangle with the singular point internal to it the procedure is the same.
On each triangle T_k we pass in local polar coordinates , , so we have

$$I_k = \int_{T_k} \frac{1}{R}\, dxdy = \int_0^{\bar{\vartheta}} d\vartheta \int_0^{\varrho(\vartheta)} \frac{1}{\varrho}\, \varrho\, d\varrho = \int_0^{\bar{\vartheta}} \varrho(\vartheta)\, d\vartheta \tag{3}$$

The parameters in equation (3) are shown in fig.2.

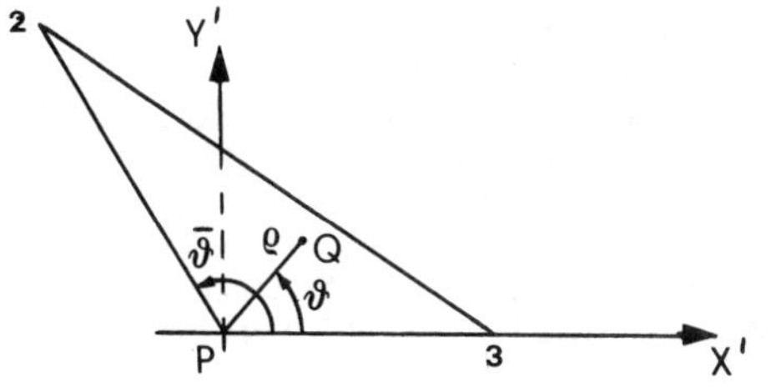

Fig.2 Local axes on a triangle and definition of parameters.

To compute the last integral in equation (3) one have to determine the equation, in polar coordinates, of the segment joining points P_3 and P_2. If m is the angular coefficient, in the local axes x', y' of segment $\overline{P_3P_2}$ it is not difficult to verify that the required equation is:

$$\varrho(\vartheta) = \frac{mx'_3}{m\cos\vartheta - \sin\vartheta} \tag{4}$$

It is to be noted that $m\cos\vartheta - \sin\vartheta$ is never zero because this would imply $\overline{P_1P_2} \,||\, \overline{P_3P_2}$.
Using equation (4) the integral (3) becomes

$$I_k = mx'_3 \int_0^{\bar{\vartheta}} \frac{d\vartheta}{m\cos\vartheta - \sin\vartheta} = \frac{2\,mx'_3}{\sqrt{1+m^2}} \left\{ \operatorname{arctanh} \frac{1+m\,\mathrm{tg}\,\frac{\bar{\vartheta}}{2}}{\sqrt{1+m^2}} - \right.$$

$$\left. - \operatorname{arctanh} \frac{1}{\sqrt{1+m^2}} \right\} \tag{5}$$

To better evaluate equation (5) by means of a computer it is useful to remember that:

$$\operatorname{arctanh}(u) = \frac{1}{2}\,\mathrm{lg}\,\frac{1+u}{1-u} \tag{6}$$

This allows to write integral (5) in the form

$$I_k = \frac{1}{\sqrt{1+m^2}}\,\mathrm{lg}\left\{ \frac{\sqrt{1+m^2}+1+m\,\mathrm{tg}\,(\bar{\vartheta}/2)}{\sqrt{1+m^2}-1-m\,\mathrm{tg}\,(\bar{\vartheta}/2)} \quad \frac{\sqrt{1+m^2}-1}{\sqrt{1+m^2}+1} \right\} \tag{7}$$

As said before it is also important to integrate, on the surface elements, the normal derivative of the fundamental function 1/R. This can be done by means of Gauss formulae when distance R is large. When source point P lies on the integration panel and we approximate the curved element by means of a plane one the integral of the normal derivative is zero so we have not to worry about it.
However it is not possible to integrate the partial derivatives of function 1/R using the technique here presented.

To check the validity of formula (7) versus Gauss formulae a simple integral over a planar triangle is presented. The geometry of the triangle is shown in fig.3

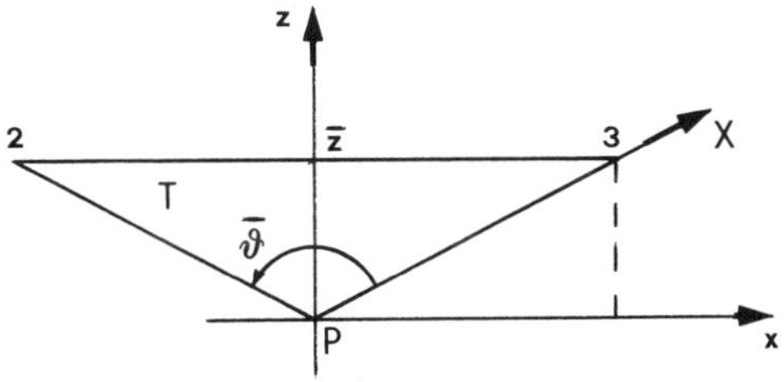

Fig.3 Simple triangle used to compute integral.

The integral over this triangle can be computed analytically by direct integration. In this case we have $\bar{\vartheta}$=153, m=-1/2, $\bar{z}$=1 and x'=$\sqrt{5}$. Results obtained using 7 Gauss points and equation (7) are as follow

exact	7 Gauss p.s	equation 7
2.88727095	3.0460288	2.88727095

It can be seen that Gauss formulae gives an error of about +5.5%. Moreover the error, using Gauss formulae, increases whenever the triangle becomes more and more irregular (i.e. when $\bar{\vartheta}$ in fig. 3 increase). In this case it is very important to integrate analytically without any error.

REFERENCES

1. Cristescu, M. and Loubignac, G., Gaussian Quadrature Formulas for Functions with Singularities in 1/r over Triangles and quadrangles, in Recent Advances in BEM, Ed. Brebbia C.A., Pentech Press 1978.

2. Aliabadi, M.H., Hall,W.S. and Phemister, T.G., Taylor Expansion for Singular Kernels in the BEM, Int. J. Num. Meth. Engng, 21, pp 2221-2236,1985.

3. Hayamy, K. and Brebbia, C.A.,Quadrature Methods for Singular and Nearly Singular Integrals in 3-D Boundary Element Method, in BEM X/88 (Ed. Brebbia C.A.), pp 237-264, Proceedings of the X Int. Conf. on BEM, Southampton 1988.

4. Newman, J.N. Marine Hydrodynamics, The MIT Press, Cambridge,Massachusetts, 1986.

5. Hess, J.L. and Smith, A.M.O., Calculation of non-Lifting Potential Flow About Arbitrary Three-Dimensional Bodies, Tec. Rep. E.S. 40622, Douglas Aircraft Company, 1962.

Numerical Integration of Kernel Shape Function Products for Singular Boundary Elements

J.Q. Zhen and D.J. Cartwright

Department of Mechanical Engineering, Bucknell University, Lewisburg, PA 17831, USA

ABSTRACT

The computing time and accuracy of numerical integration schemes for kernel-shape function products over singular boundary elements has been investigated. The methods compared are element subdivision, a cubic polynomial transformation and a quintic polynomial transformation. The comparisons have been carried out for three dimensional elasto-statics using a cube subjected to uniaxial linearly varying tractions. It is shown that, for similar computing time, the element subdivision method is within 0.01% of known accurate results whereas both the cubic and quintic polynomial transformations give accuracies of 4% and 0.5% respectively. To obtain an accuracy comparable with the subdivision method the quintic polynomial transformation requires at least two and a half times as much computer time and the cubic polynomial transformation requires even more time. It is shown that this increase in time requirement for the two polynomial transformation methods occurs because of the increased number of Gauss integration points and numerical operations needed.

INTRODUCTION

The solution of boundary integral equations by the boundary element method (BEM) involves numerically evaluating integrals over each of the elements. These numerical integrations take a major part of the CPU time of any BEM computer solution. There is thus a need to develop efficient integration procedures. A further difficulty arises because of the need to evaluate singular integrals which arise when the source point is located on the element over which the integration is performed. These terms arising from the singular integrals lie on the leading diagonals of the BEM matrices and therefore have a strong effect on the solution.

For constant and linear elements of two-dimensional and three-dimensional potential and elasto-static fields these singular integrals have been evaluated analytically [1-4]. For two-dimensional quadratic isoparametric elements the kernel-shape function products which make up the integrand of the singular integral may be changed into a non-singular integrand together with an integrand with a logarithmic singularity. These are evaluated numerically using a method based on logarithmic Gaussian quadrature [5]. For three-dimensional fields, the singular integrals have been calculated by subdividing a quadratic, quadrilateral element into two or three triangular elements, depending on the position of the source point, (i.e. corner or mid-side node) and using a linear intrinsic coordinate transformation prior to Gaussian quadrature [6]. This subdivision method has been successful in three-dimensional BEM computer programs but there is a need to develop a method which has a simpler algorithm, in order to avoid tedious programming, and which will save computing time and further improve the accuracy of the boundary element method. A coordinate transformation has been proposed [7] in which the singularity strength in the integrand is reduced. This method has been successfully applied to curved boundary elements but sub-division of the element is still required. A polynomial coordinate transformation has been proposed [8] which attempts to eliminate the singularity in the integrands. This leads to a simpler algorithm but the third degree polynomial transformation used [8] was insufficient to deal accurately with the singular integrand. An alternative method is to use a Taylor series expansion of the singular integral kernels at the source point [9] and divide the original singular integral into a new singular integral, that can be calculated analytically, and a non-singular part which can be evaluated by Gaussian quadrature.

The polynomial coordinate transformation method [8] leads to a relatively simple algorithm and therefore would be useful if its accuracy could be improved. In the present paper the degree of the coordinate transformation is increased in an attempt to improve accuracy of the method and to evaluate it for general use in three-dimensional boundary element programs.

THEORETICAL FORMULATION

The solution of boundary element equations in 3-D typically require numerical evaluation of integrals I of the form

$$I = \int_{-1}^{+1}\int_{-1}^{+1} \frac{N_\alpha(\bar{\xi})J(\bar{\xi})}{r(\bar{\xi},\bar{\xi}^s)} \, d\xi_1 \, d\xi_2 \qquad (1)$$

where $\bar{\xi} = (\xi_1,\xi_2)$ is the intrinsic coordinate, $J(\bar{\xi})$ the Jacobian, $N_\alpha(\bar{\xi})$ a two-degree polynomial shape function, and $r\,(\bar{\xi}, \bar{\xi}^s)$ the distance between a field point of intrinsic coordinate $\bar{\xi}=(\xi_1, \xi_2)$ and a source point of intrinsic coordinate $\bar{\xi}^s = (\xi_1^s, \xi_1^s)$ given by

$$r(\bar{\xi},\bar{\xi}^s) = \sqrt{\sum_{i=1}^{3} \{x_i^\alpha \, [N_\alpha(\bar{\xi}^s) - N_\alpha(\xi)]\}^2} \qquad (2)$$

Thus r becomes an infinitesimal quantity of degree two as the field point approaches the source point. This gives rise to a singularity of order two in the integrand of equation (1). If the Jacobian can be expressed to cancel this singularity as $\bar{\xi} \to \bar{\xi}^s$ then a non-singular integral will result.

This can be done by introducing a polynomial transformation between the intrinsic coordinates (ξ_1, ξ_2) and new coordinates $\bar{\eta} = (\eta_1,\eta_2)$ as

$$\xi_i(\eta_i) = \sum_{j=0}^{N} a_{ij}^s \, \eta_i^j \qquad i=1,2 \qquad (3)$$

where the constants a_{ij}^s, which depend on the position of the source points, and the order N are determined from specified conditions for the transformation. Using this formulation the integral in equation (1) becomes

$$I = \int_{-1}^{+1}\int_{-1}^{+1} \frac{N_\alpha[\bar{\xi}(\bar{\eta})]J[\bar{\xi}(\bar{\eta})]J_o(\bar{\eta})}{r\,[\bar{\xi}(\bar{\eta}),\bar{\xi}(\bar{\eta}^s)]} \, d\eta_1 \, d\eta_2 \qquad (4)$$

where the additional Jacobian $J_o(\bar{\eta})$ is given by

$$J_o(\bar{\eta}) = \frac{d\xi_1}{d\eta_1}\frac{d\xi_2}{d\eta_2} \qquad (5)$$

By a suitable choice of the polynomial (3), the additional Jacobian $J_o(\bar{\eta})$ can be arranged to cancel the singularity in the integrand which can then be evaluated using Gauss quadrature as

$$I = \sum_{k=1}^{K} \sum_{\ell=1}^{L} W_k W_\ell \left\{\frac{N_\alpha[\bar{\xi}(\eta_{1k},\eta_{2\ell})]J[\bar{\xi}(\eta_{1k},\eta_{2\ell})]J_o(\eta_{1k},\eta_{2\ell})}{r\,[\bar{\xi}(\eta_{1k},\eta_{2\ell}),\ \bar{\xi}(\eta_1^s,\eta_2^s)]}\right\} \qquad (6)$$

where η_{1k} and $\eta_{2\ell}$ are the transformed positions of Gauss integral points, W_k and W_ℓ are the weights and $(\eta_1^s, \eta_2^s) = \bar{\eta}^s$ is the transformed position of the source point.

A third degree transformation and a fifth degree transformation are used in the present work, however since the third degree has been treated elsewhere [8] the derivation for a fifth degree only will be carried out. For this case equation (3) becomes

$$\xi_i = \sum_{j=o}^{5} a_{ij}^s\, \eta_i^j \qquad i=1,2 \qquad (7)$$

The requirement that the square $\{-1\le\xi_1\le+1;\ -1\le\xi_2\le+1\}$ in the (ξ_1,ξ_2) coordinate system be transformed to another square $\{-1\le\eta_1\le+1;\ -1\le\eta_2\le+1\}$ in (η_1,η_2) coordinate system, so that Gauss quadrature can be used, is satisfied by ensuring

$$\xi_i(1) = 1 \text{ and } \xi_i(-1) = -1, \text{ for } i = 1,2 \qquad (8)$$

The requirement that the singularity in the integrand of (4) be completely eliminated is given by

$$\left.\frac{d^m\xi_i(\eta_i)}{d\eta_i^m}\right|_{\eta_i^s} = o, \text{ for } m = 1,2,3,4; \quad i = 1,2 \qquad (9)$$

where η_i^s is the coordinate of the source point in the η_i coordinate system. Using equations (8) and (9) gives two systems of 6 equations which can be solved for the unknown constants a_{ij}^s which are given in appendix A. Hence from equation (5) the additional Jacobian J_0 is given by

$$J_0 = (5a_{15}\eta_1^{s\,4} + 4a_{14}\eta_1^{s\,3} + 3a_{13}\eta_1^{s\,2} + 2a_{12}\eta_1^{s} + a_{11}^{s}) \qquad (10)$$

$$(5a_{25}\eta_2^{s\,4} + 4a_{24}\eta_2^{s\,3} + 3a_{23}\eta_2^{s\,2} + 2a_{22}\eta_2^{s} + a_{21}^{s})$$

$$s = 1,8.$$

This now allows the singularity to be removed from the integrand of I in equation (4).

MODEL STUDY

The polynomial transformation has been programmed for polynomials up to degree five and the coefficients of each transformation determined. The integration procedures based on this transformation have been introduced into a three-dimensional elasto-static boundary element program. The accuracy of the method, and the execution time, have been studied using the model configuration shown in fig. 1. This consists of a cube of dimension 10 X 10 X 10 (cm) subjected to a uniaxial bending traction of magnitude $100N/(cm)^2$ on the face $x_1 = 10$ and restrained on the face $x_1 = o$ as shown. The cube is modelled by 6 boundary elements and 20 nodes as shown in figure 1.

RESULTS AND DISCUSSION

The results in table 1a show the CPU time for calculation of the singular integral for a fifth degree polynomial transformation. It is shown that when the source point is at a corner node the subdivision method requires less time for the calculation of the singular integral over the element (1) than when the source point is a mid-side node (0.28 millisec compared to 0.43 millisec). In contrast the computation time for the fifth degree polynomial transformation is insensitive to the position of the source point. However the computation time for the transformation method increases significantly with the number of Gauss integration points. In table 1b the, entire computing time of the program, and the accuracy, of the x_1 direct stress component are shown for the fifth degree polynomial transformation and the element subdivision method. It can be seen that, for

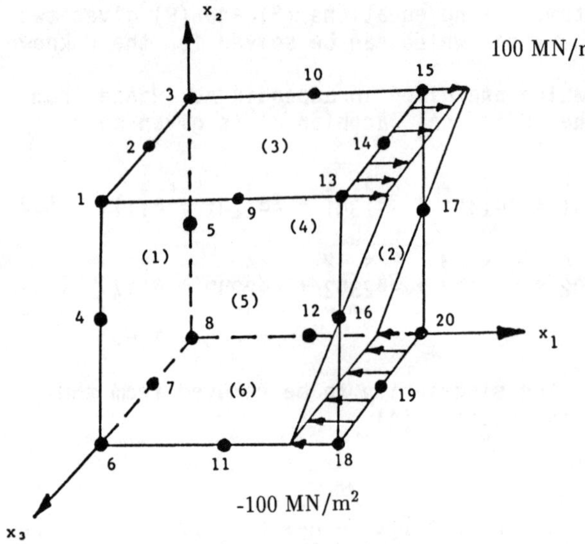

node	Component u_1	u_2	u_3
1	0	-	-
2	0	0	0
3	0	-	-
4	0	-	-
5	0	-	-
6	0	-	-
7	0	-	0
8	0	-	-

Element 1 Constraints

- unconstrained

0 zero

Fig. 1 Cube subjected to a uniaxial bending traction.

Number of Gauss Points	$\iint U_{11} N_1 ds$ S_1 Source at Corner-node	CPU Millesec.	$\iint U_{11} N_5 ds$ S_1 Source at Mid-side node	CPU Millesec.
36	0.576905E-01 0.000000E+00 0.000000E+00	0.17	0.642119E+00 0.000000E+00 0.000000E+00	0.17
813	0.549845E-01 0.000000E+00 0.000000E+00	0.39	0.646751E+00 0.000000E+00 0.000000E+00	0.39
144	0.545717E-01 0.000000E+00 0.000000E+00	0.68	0.646225E+00 0.000000E+00 0.000000E+00	0.68
225	0.544884E-01 0.000000E+00 0.000000E+00	1.17	0.646203E+00 0.000000E+00 0.000000E+00	1.16
element subdiv. 36	0.544594E-01 0.000000E+00 0.000000E+00	0.28	0.646211E+00 0.000000E+00 0.000000E+00	0.43

Table 1a. CPU time of singular integrals for the fifth degree transformation

Node	Number of Gauss Points 36	81	144	225	Element Subdivision 36
1	97.07	100.06	99.96	99.99	100.00
2	97.81	100.05	99.96	99.99	100.00
3	97.07	100.06	99.96	99.99	100.00
4	0.33	0.00	0.00	0.00	0.00
5	0.33	0.00	0.00	0.00	0.00
6	- 96.65	-100.06	- 99.95	- 99.98	-100.00
7	- 98.11	-100.05	- 99.96	- 99.99	-100.00
8	- 96.65	-100.06	- 99.95	- 99.98	-100.00
9	100.52	100.30	100.05	100.01	100.00
10	100.52	100.30	100.05	100.01	100.00
11	-100.46	-100.30	-100.05	-100.01	-100.00
12	-100.46	-100.30	-100.05	-100.01	-100.00
13	102.42	100.26	100.07	100.02	100.00
14	101.25	100.14	100.02	100.01	100.00
15	102.42	100.26	100.07	100.01	100.00
16	0.04	0.00	0.00	0.00	0.00
17	0.04	0.00	0.00	0.00	0.00
18	-102.43	-100.26	-100.07	-100.02	-100.00
19	-101.32	-100.14	-100.02	-100.01	-100.00
20	-102.43	-100.26	-100.07	-100.02	-100.00
err. %	4	0.5	0.07	0.02	0.005
CPU Time Sec.	22.0	45.0	76.0	117.0	45.0

Table 1b. CPU time and accuracy of the fifth degree transformation for the direct stress component in the x_1 direction

similar computing time, the element subdivision method is within 0.005% of the known accurate results whereas the fifth degree transformation is only within 0.5%. To obtain an accuracy that is comparable with the subdivision method the fifth degree polynomial transformation requires more than two and a half times as much computing time. This increase results from the large number of Gauss integration points; i.e. 225 for the fifth degree polynomial transformation compared with 36 for the subdivision method. Similar results are presented in tables 2a and 2b for the third-degree polynomial as proposed by Telles [8]. It can be seen from table 2a that the computing times for the transformation method are again insensitive to the position of the source point and that, as before, times increase significantly with the number of Gauss points. In Table 2b it is shown that the third-degree polynomial transformation does not give comparable accuracy to the element subdivision method even when the number of Gauss points is increased to 225 and furthermore the accuracy remains about two orders of magnitude less than the element subdivision method. Also the time penalty is significant in that even for this lower accuracy the third-degree polynomial transformation requires two and a half times the computing time of that required by element subdivision.

A further comparison can be made between the fifth degree and third degree transformation from tables 1b and 2b. It can be seen that for similar cpu time and the same number (225) of Gauss points the third degree transformation is considerably less accurate than the fifth degree (0.7% compared to 0.02% error). This occurs because the Jacobian J_0 arising from the third degree polynomial does not entirely remove the singularity in the integrand of equation (4).

Number of Gauss Points	$\int\int U_{11}N_1 ds$ S_1 Source at Corner-node	CPU Millesec.	$\int\int U_{11}N_5 ds$ S_1 Source at Mid-side node	CPU Millesec.
36	0.567241E-01 0.000000E+00 0.000000E+00	0.16	0.662371E+00 0.000000E+00 0.000000E+00	0.16
81	0.548932E-01 0.000000E+00 0.000000E+00	0.36	0.640620E+00 0.000000E+00 0.000000E+00	0.36
144	0.545737E-01 0.000000E+00 0.000000E+00	0.84	0.647981E+00 0.000000E+00 0.000000E+00	0.82
225	0.544969E-01 0.000000E+00 0.000000E+00	1.10	0.645038E+00 0.000000E+00 0.000000E+00	1.00
element subdiv. 36	0.544594E-01 0.000000E+00 0.000000E+00	0.28	0.646211E+00 0.000000E+00 0.000000E+00	0.43

Table 2a. CPU time of singular integrals for the third degree transformation

Node	Number of Gauss Points 36	81	144	225	Element Subdivision 36
1	106.67	97.20	100.81	99.43	100.00
2	106.51	97.03	100.92	99.38	100.00
3	106.67	97.20	100.81	99.43	100.00
4	- 0.61	0.28	- 0.08	0.06	0.00
5	- 0.61	0.28	- 0.08	0.06	0.00
6	-107.51	- 96.84	-100.91	- 99.35	-100.00
7	-106.02	- 97.27	-100.85	- 99.43	-100.00
8	-107.51	- 96.84	-100.91	- 99.35	-100.00
9	103.52	99.39	100.28	99.85	100.00
10	103.52	99.39	100.28	99.85	100.00
11	-103.64	- 99.34	-100.29	- 99.84	-100.00
12	-103.64	- 99.34	-100.29	- 99.84	-100.00
13	99.64	100.90	99.85	100.16	100.00
14	100.76	99.87	100.13	99.95	100.00
15	99.64	100.90	99.85	100.16	100.00
16	0.08	0.03	- 0.01	0.01	0.00
17	- 0.08	0.03	0.01	0.01	0.00
18	- 99.63	-100.91	- 99.85	-100.16	-100.00
19	-100.62	- 99.93	-100.11	- 99.97	-100.00
20	- 99.63	-100.91	- 99.85	-100.16	-100.00
err. %	8	4	1	0.7	0.005
CPU Time Sec.	22	43	73	111	45

Table 2b. CPU time and accuracy of the third degree transformation for the direct stress component in the x_1 direction

CONCLUSIONS

1. The algorithm for the polynomial coordinate transformation method is simpler to program than that for the element subdivision method.

2. For comparable accuracy the computing time of the element subdivision method is significantly less than that required for the polynomial coordinate transformation method.

3. The accuracy of the polynomial transformation method increases with the polynomial degree and with the number of Gauss integration points.

4. At least 225 Gauss integration points and a fifth degree polynomial transformation are required to obtain an accuracy comparable to the element subdivision method.

5. At comparable computing times the third degree polynomial transformation is less accurate than the fifth degree, for integration over the singular element, since it does not entirely remove the singularity.

ACKNOWLEDGEMENTS

The authors wish to thank Professor Dan Hyde for his assistance with coding the cpu clock timing, Gloria Auman for typing the manuscript and the financial support of the Ben Franklin program together with Swanson Analysis Systems Inc.

REFERENCES

1. Cruse, T.A., Numerical Solution in Three Dimensional Elastostatics, Int. J. solids structures, Vol. 5, pp. 1259-1274, 1969.

2. Brebbia, C.A.,The Boundary Element Method for Engineers, Pentech Press, London, 1978.

3. Banerjee, P.K. and Butterfield R., Boundary Element Methods in Engineering Science, McGraw-Hill Book Company, 1981.

4. Cruse, T.A., Swedlow, J.L., Interactive Program for Analysis and Design Problems in Advanced Composite Technology, Technical Report AFML-TR-71-168, 1971.

5. Stroud, A.H. and Secrest, D., Gaussian Quadrature Formulas, Prentice-Hall, New York, 1966.

6. Lachat, J.C. and Watson, J.O., Effective Numerical Treatment of Boundary Integral Equations: A Formulation for Three Dimensional Elastostatics, Int. J. Num. Meth. Eng., Vol. 10, pp. 991-1005, 1976.

7. Hayami, K. and Brebbia, C.A., A New Coordinate Transformation Method for Singular and Nearly Singular Integrals over General Curved Boundary Elements, In Boundary Elements IX, Vol. 1, pp. 375-399, edited by Brebbia, C.A., Wendland, W.L. and Kuhn, G., Springer-Verlag, 1987.

8. Telles, J.C.F., A Self-adaptive Co-ordinate Transformation for Efficient Numerical Evaluating of General Boundary Element Integrals, Int. J. Num. Meth. Eng., Vol. 24, pp. 959-973, 1987.

9. Aliabadi, M.H., and Hall, W.S. Taylor Expansions for Singular Kernels in the Boundary Element Method, Int. J. Num. Method in Eng., Vol. 21, pp. 2221-2236, 1985.

Appendix A

Coefficients of Quintic Polynomial Transformation

Position of source points

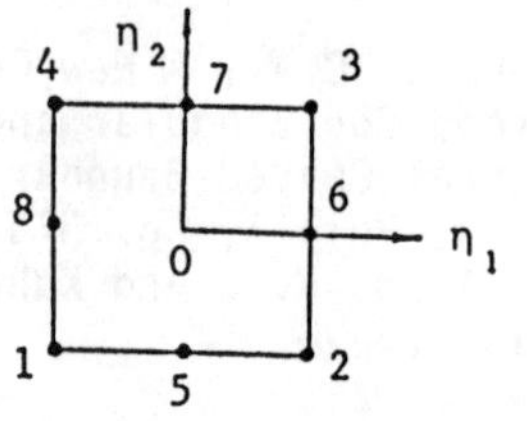

Fig·A1 Location of source point S in tables A1 and A2

s \ j	1	2	3	4	5	6
1	0.0625	0.3125	0.6250	0.6250	0.3125	-0.9375
2	0.0625	-0.3125	0.6250	-0.6250	0.3125	0.9375
3	0.0625	-0.3125	0.6250	-0.6250	0.3125	0.9375
4	0.0625	0.3125	0.6250	0.6250	0.3125	-0.9375
5	1.0000	0.0000	0.0000	0.0000	0.0000	0.0000
6	0.0625	-0.3125	0.6250	-0.6250	0.3125	0.9375
7	1.0000	0.0000	0.0000	0.0000	0.0000	0.0000
8	0.0625	0.3125	0.6250	0.6250	0.3125	-0.9375

Table A1. Coefficients a_{1j}^{s}

s \ j	1	2	3	4	5	6
1	0.0625	0.3125	0.6250	0.6250	0.3125	-0.9375
2	0.0625	0.3125	0.6250	0.6250	0.3125	-0.9375
3	0.0625	-0.3125	0.6250	-0.6250	0.3125	0.9375
4	0.0625	-0.3125	0.6250	-0.6250	0.3125	0.9375
5	0.0625	0.3125	0.6250	0.6250	0.3125	-0.9375
6	1.0000	0.0000	0.0000	0.0000	0.0000	0.0000
7	0.0625	-0.3125	0.6250	-0.6250	0.3125	0.9375
8	1.0000	0.0000	0.0000	0.0000	0.0000	0.0000

Table A2. Coefficients a_{2j}^{s}

Treatment of Corner Node Problems in Boundary Element Method

Geng Yan* and Feng-Bao Lin**

*Department of Civil Engineering, Chongqing Institute of Architecture and Engineering, Chongqing, China

(currently visiting the Polytechnic University, Brooklyn).

**Department of Civil and Environmental Engineering, Polytechnic University, Brooklyn, NY 11201, USA

ABSTRACT

The geometric corner nodes in which the outward normals and the tractions are discontinuous need a special treatment in a boundary element analysis. Whether the problems are treated properly directly influences the accuracy of calculation results. This paper presents a systematical review of the treatments of the corner node problems. Three types of treatments are classified: (1) methods with continuous boundary displacements and tractions; (2) methods with discontinuous boundary displacements and tractions; and (3) methods with continuous boundary displacements and discontinuous tractions. The boundary point element method is then proposed. A simple formula to calculate boundary stresses at corner nodes is given. Also a method to deal with the singularity due to displacement boundary conditions at a boundary point element is presented. Several typical examples in the treatment of the corner node problems using different methods are carried out and their performances are compared.

INTRODUCTION

The boundary integral equation for elasticity problem is [3, 13]

$$C_{ij}(P)u_j(P) + \int_\Gamma u_j(Q)T_{ij}(P,Q)d\Gamma(Q) =$$

$$\int_\Gamma \tau_j(Q)U_{ij}(P,Q)\ d\Gamma(Q) + \int_\Omega b_j(q)U_{ij}(P,q)\ d\Omega(q) \qquad (1)$$

where u_j, τ_j, and b_j are displacements, tractions, and body forces; U_{ij} and T_{ij} are coefficients related to geometric positions of P and Q (or q) and elastic parameters of the body; $C_{ij}(P)$ is a constant depending on the geometric conditions of boundary point P. If P is at a smooth boundary, $C_{ij}(P) = \delta_{ij}/2$.

In the numerical implementation, the body is divided into M boundary elements Γ_e, e = 1,2,...,M, with N boundary nodes. The boundary quantities can then be expressed through the interpolation functions at nodes. Thus at each boundary node, Eq.(1) becomes [13]

$$\sum_{m=1}^{N} \tilde{T}_{ij}^{nm} u_j^m = \sum_{m=1}^{N} \tilde{U}_{ij}^{mn} \tau_j^m \quad (n = 1,2,...,N) \tag{2}$$

where u_j^m and τ_j^m are values of u_j and τ_j at node Q_m, and

$$\tilde{T}_{ij}^{nm} = \sum_{e=1}^{M} \int_{\Gamma_e} T_{ij}(P_n,Q)\, N^m d\Gamma + C_{ij}(P_n)\delta_{nm} \tag{3}$$

$$\tilde{U}_{ij}^{nm} = \sum_{e=1}^{M} \int_{\Gamma_e} U_{ij}(P_n,Q)\, N^m d\Gamma \tag{4}$$

in which N^m is the shape function of the mth node.

For any given boundary conditions, Eq.(2) can always be transformed into a linear algebraic system of equations

$$\underset{\sim}{A}\, \underset{\sim}{x} = \underset{\sim}{b} \tag{5}$$

The calculations of coefficients T_{ij}^{nm} and U_{ij}^{nm}, which are related to the interpolation functions and the type of boundary elements, are obviously the key issues in constructing Eq. (5). For a two-dimensional problem, if non-constant elements, such as linear or quadratic elements, are chosen, the boundary nodes may locate at non-smooth parts of the boundary. These boundary nodes, called corner nodes, cause some special problems: (1) The relation of $C_{ij} = \delta_{ij}/2$ is no longer valid and should be calculated through different procedures; (2) The direction of the boundary normal at a corner node is discontinuous, so are the tractions at the node. In this case, it is necessary to distinguish two different tractions, each one located on either side of the node.

Several methods have been proposed to treat the problems. They can generally be classified into three types:

1) Methods with continuous boundary displacements and tractions [1, 7].

2) Methods with discontinuous boundary displacements and tractions, such as corner node splitting method [1, 7], composite boundary element method [11], and discontinuous boundary element method [7, 8].

3) Methods with continuous boundary displacements and discontinuous boundary tractions such as boundary point element method.

The paper reviews and studies above methods. The boundary point element method is then presented. A simple formula to calculate boundary stresses at corner nodes is given. Also a method to deal with the singularity problem that occurs when the displacements are known at both nodes of a boundary point element is proposed. Several examples are carried out to compare the performances of the methods.

METHODS WITH CONTINUOUS BOUNDARY DISPLACEMENTS AND TRACTIONS

The influence of discontinuity of normal directions at a corner node is neglected in this method and the tractions on both sides of the node are assumed equal.

In this case the problem of determining $C_{ij}(P_n)$ still exists. Two ways are generally used, one is the analytical method [6] and the other is rigid-body displacement method [2, 4]. The latter method is considered more convenient in which a unit rigid displacement along certain direction of coordinate is assumed. Then from Eq.(2), we have

$$\tilde{\tilde{T}}_{ij}^{nn} = -\sum_{\substack{m=1 \\ m\neq n}}^{N} \tilde{\tilde{T}}_{ij}^{nm}, \quad (i,j = 1,2;\ n = 1,2,\ldots,N) \qquad (6)$$

Because the term $C_{ij}(P)$ exists only when n = m, i.e., $C_{ij}(P)$ is included in the $\tilde{\tilde{T}}_{ij}^{nn}$ which can be directly calculated from Eq.(6).

METHODS WITH DISCONTINUOUS BOUNDARY DISPLACEMENTS AND TRACTIONS

Corner node splitting method

A corner node is split into two nodes, each one is then moved a slight distance δ along either direction of the two adjacent boundary elements at the corner. Since the split nodes are no longer located at a corner, the value of C_{ij} is just $\delta_{ij}/2$ at either node. The tractions at the two nodes can represent the

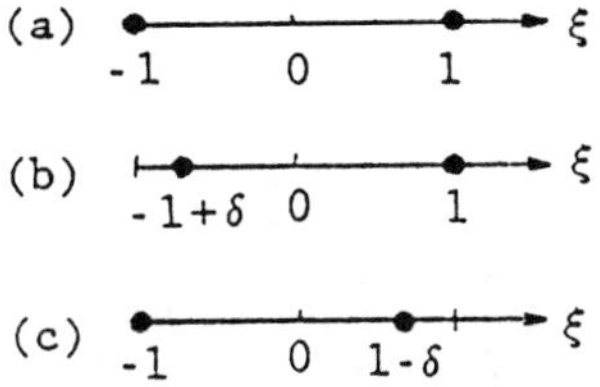

Fig.1

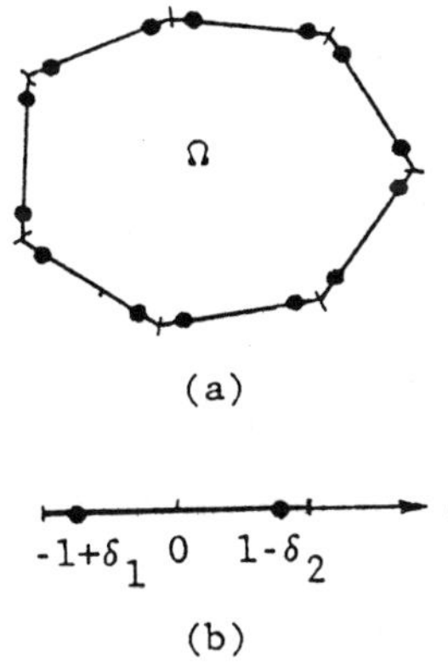

Fig.2

discontinuity at the corner. The shape functions in this case must be reconstructed. For the linear element shown in Fig. 1b,

$$N_1 = \frac{1}{2-\delta}(1 - \xi), \quad N_2 = \frac{1}{2-\delta}(1 - \delta + \xi); \tag{7}$$

for the one shown in Fig. 1c,

$$N_1 = \frac{1}{2-\delta}(1 - \delta - \xi), \quad N_2 = \frac{1}{2-\delta}(1 + \xi). \tag{8}$$

Discontinuous boundary elements methods

The terminal nodes of all elements including nodes at and not at corners are moved inside the elements (Fig. 2a). Thus both tractions and displacements of neighbouring elements are discontinuous and the corner node problems disappear.

The shape functions adopted in this method for a linear element are

$$N_1 = \frac{1}{2-\delta_1-\delta_2}(1 - \delta_2 - \xi), \quad N_2 = \frac{1}{2-\delta_1-\delta_2}(1 - \delta_1 + \xi) \tag{9}$$

By properly adjusting the values of δ_1 and δ_2, different discontinuous boundary elements are obtained. Some researchers indicated that the results with higher accuracy can be achieved when the nodes are chosen at Gauss integration points [8]. The drawback of the method is the increase of the total degrees of freedom.

Composite boundary elements method

In the method, different boundary elements are used at the same time in order to achieve both the advantage of higher

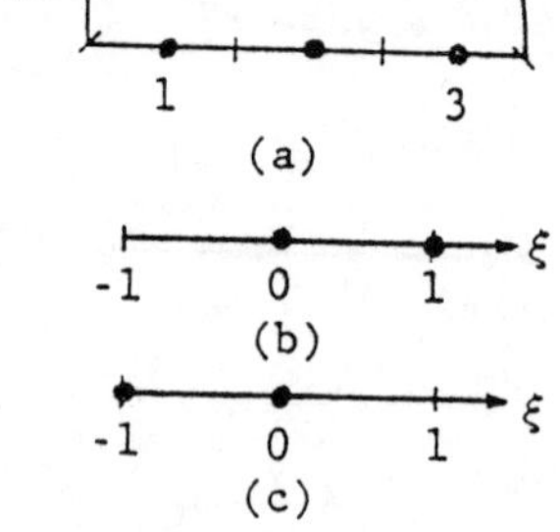

Fig.3

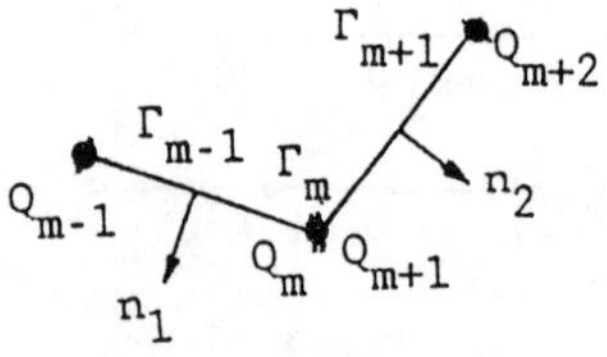

Fig.4

order elements and the avoidance of the corner node problems. Fig. 3a shows the example of the composite boundary elements which combines the constant and linear elements. The constant elements are used near corners in order to prevent nodes from locating at corners.

Two kinds of link-elements are needed to connect linear elements with constant elements. For the link-element as element 4 in Fig. 3b, the shape functions are $N_1(\xi) = 1 - \xi$, $N_2(\xi) = \xi$; and for the link-element as element 7 in Fig. 3c, the shape functions are $N_1(\xi) = -\xi$, $N_2(\xi) = 1 + \xi$.

Similarly, other types of composite boundary elements can be constructed for different cases.

METHODS WITH CONTINUOUS BOUNDARY DISPLACEMENTS AND DISCONTINUOUS TRACTIONS

The displacements at a boundary corner are always continuous but the tractions could be continuous or discontinuous depending on the given boundary conditions. The boundary point element method is developed to satisfy these actual situations.

A boundary point element established at each corner is a special kind of linear boundary element whose length is zero and two nodes are located at the same geometric position. The element is depicted in Fig. 4 where Γ_m is a boundary point element with two nodes, Q_m and Q_{m+1}, located at the same position. Owing to two nodes at a corner, the element can simulate both the continuous displacements and discontinuous tractions.

STRESSES AT A CORNER POINT

The stresses at a corner point can not be calculated from the regular procedure. The difficulty can easily be solved by the boundary point element method. The stresses σ_x, σ_y, and τ_{xy} at the two nodes of a boundary point element must be the same since the stress state at any point within an elastic body is unique. From the relation of $\tau_j = \sigma_{ij} n_i$ for the two nodes of a boundary point element, we have

$$
\begin{aligned}
l_1\sigma_x + m_1\tau_{xy} &= \tau_x^m, \qquad & m_1\sigma_y + l_1\tau_{xy} &= \tau_y^m \\
l_2\sigma_x + m_2\tau_{xy} &= \tau_x^{m+1}, \qquad & m_2\sigma_y + l_2\tau_{xy} &= \tau_y^{m+1}
\end{aligned}
\tag{10}
$$

where (l_1, m_1) and (l_2, m_2) are the direction cosines of outward normals to elements Γ_{m-1} and Γ_{m+1} (Fig. 4). τ_x^m, τ_y^m and τ_x^{m+1}, τ_y^{m+1} are tractions at two nodes Q_m and Q_{m+1} respectively. Using any three of Eq.(10), such as the first three equations, we get

$$\sigma_x = \frac{1}{\alpha}(m_2\tau_x^m - m_1\tau_x^{m+1})$$
$$\sigma_y = \frac{1}{\alpha}(l_2\tau_x^m - l_1\tau_x^{m+1})\frac{l_1}{m_1} + \frac{\tau_y}{m_1} \qquad (11)$$
$$\tau_{xy} = \frac{1}{\alpha}(l_1\tau_x^{m+1} - l_2^m\tau_x^m)$$

where $\alpha = l_1m_2 - m_1l_2$

The conditions of using Eq.(11) are $l_1/m_1 \neq l_2/m_2$ and $m_1 \neq 0$. If $l_1/m_1 = l_2/m_2$, the two outward normals to elements Γ_m and Γ_{m+1} are parallel. This means that the boundary is smooth at nodes Q_m and Q_{m+1} so the ordinary way of calculating stresses can be applied. If $m_1 = 0$ (l_1 must be either 1 or -1), alternate formula can be found from the first two and the last equations of Eq.(10) as

$$\sigma_x = \tau_x^m / l_1$$
$$\sigma_y = (l_1\tau_y^{m+1} - l_2\tau_y^m)/(l_1m_2) \qquad (12)$$
$$\tau_{xy} = \tau_y^m / l_1$$

SINGULARITY OF BOUNDARY POINT ELEMENT

When the displacement along one direction or two directions at both nodes of a boundary point element is prescribed, the matrix $\underset{\sim}{A}$ in Eq.(5) becomes singular. To cope with the singularity, we keep the two equations in Eq.(5) corresponding to either of the above two nodes but replace the two equations corresponding to the other node by two complementary equations.

The first complementary equation is obtained by substitution of Eq.(11) into the last equation of (10) as

$$l_2\tau_x^m + m_2\tau_y^m - l_1\tau_x^{m+1} - m_1\tau_y^{m+1} = 0 \qquad (13)$$

To obtain the second complementary equation, first from the Hook's law and Eq.(10), we get

$$m_2\tau_x^m - l_2\tau_y^m - m_1\tau_x^{m+1} + l_1\tau_y^{m+1} = 2\alpha(\lambda + \mu)(\epsilon_x + \epsilon_y) \qquad (14)$$

in which ϵ_x and ϵ_y are strains at boundary point element Γ_m. λ and μ are Lame's constants. Then from the strain-displacement relationship, transformation of coordinates, and the interpolation functions for linear elements, we have

$$\epsilon_x = \frac{1}{\alpha}\left(l_2\frac{u_x^m - u_x^{m-1}}{2d_{m-1}} - l_1\frac{u_x^{m+2} - u_x^m}{2d_{m+1}}\right)$$

$$\epsilon_y = \frac{1}{\alpha}\left(m_2 \frac{u_y^m - u_y^{m-1}}{2d_{m-1}} - m_1 \frac{u_y^{m+2} - u_y^m}{2d_{m+1}}\right) \qquad (15)$$

in which $2d_{m-1}$ and $2d_{m+1}$ are the lengths of elements Γ_{m-1} and Γ_{m+1}; u_i^{m-1}, u_i^m, and u_i^{m+2} are the displacements at nodes Q_{m-1}, Q_m, and Q_{m+2} (Fig. 4). u_i^m equals u_i^{m+1} always. Substitution of Eq.(15) into Eq.(14) yields the second complementary equation as

$$m_2\tau_x^m - l_2\tau_y^m - m_1\tau_x^{m+1} + l_1\tau_y^{m+1} = a - b_x u_x^{m-1} - b_y u_y^{m-1} - c_x u_x^{m+2} - c_y u_y^{m+2} \qquad (16)$$

where $a = (b_x + c_x)\bar{u}_x^m + (b_y + c_y)\bar{u}_y^m$

$$b_x = (\lambda + \mu)l_2/d_{m-1}, \quad b_y = (\lambda + \mu)m_2/d_{m-1}$$

$$c_x = (\lambda + \mu)l_1/d_{m+1}, \quad c_y = (\lambda + \mu)m_1/d_{m+1} \qquad (17)$$

in which $\bar{u}_x^m$ and $\bar{u}_y^m$ are prescribed displacements at node Q_m of the boundary point element.

NUMERICAL EXAMPLES

Steady state heat conduction in a square plate

A 6 x 6 square plate with four corner nodes O, A, B, and C is shown in Fig. 5. The treatments of corner nodes discussed previously can also be applied to the heat conduction problems. The boundary conditions of the plate are shown in Fig. 5. Linear elements are used in the analyses. For comparison, the results of temperature u at some typical points are listed in table 1. Fig. 6 shows the results of temperature along OA and the heat flux along AC. In them, (a) shows the results of the method with continuous boundary variables (MB = NB = 16); (b) corner node splitting method (MB = 12, NB= 16); (c) discontinuous element method (MB = 12, NB= 24); (d) composite boundary element method using constant and linear elements (MB = 12, NB = 16); and (e) boundary point element method (MB = NB = 8). MB and NB are the total number of elements and nodes.

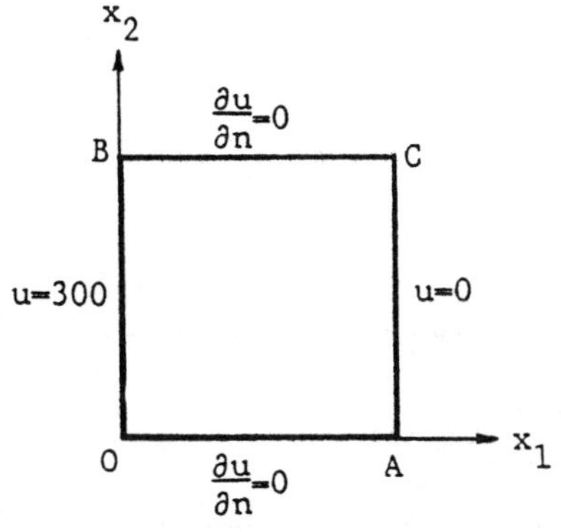

Fig.5

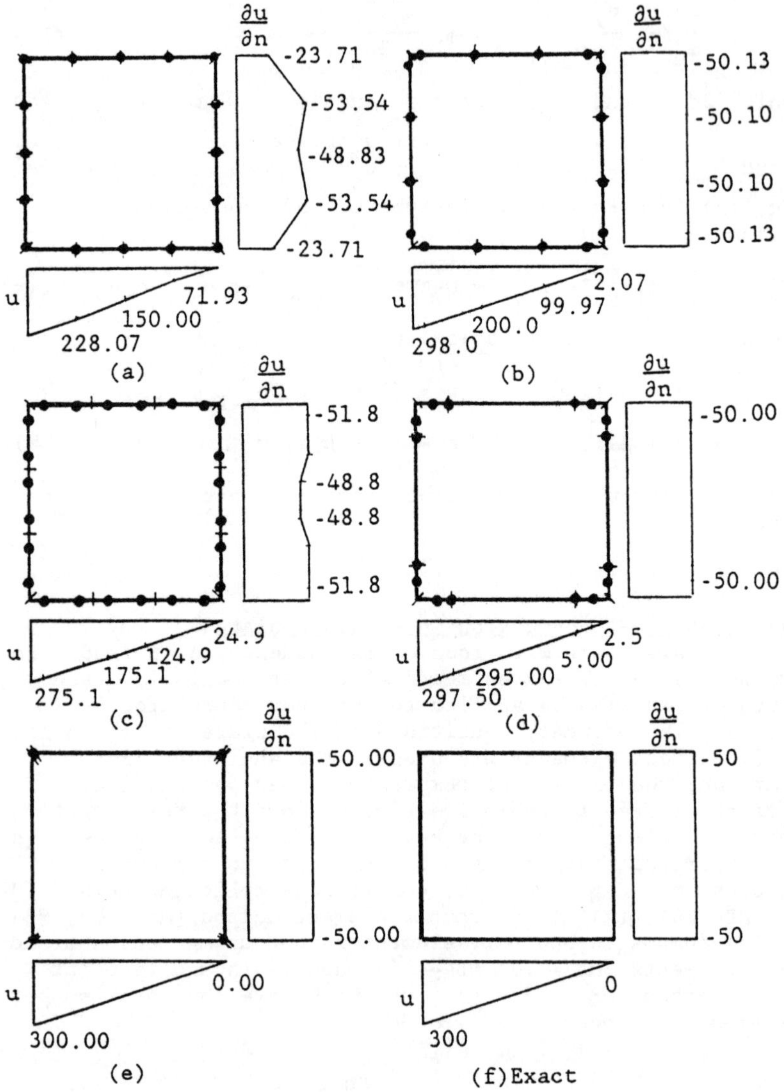

Fig.6

Table 1

Point x	y	(a)	(b)	(d)	(e)	Exact
2	2	200.325	200.0	200.000	200.000	200
3	3	150.000	150.0	150.000	150.000	150
4	4	99.675	100.0	100.000	100.000	100

Steady state heat conduction in a circular plate with a hole

The inside radius of the plate is 1 cm and the outside radius is 5 cm. A uniform temperature of 20 °C is prescribed along the external surface and a flux of 31.21 °C/cm is given across the center circular hole. Only a quarter of the plate is considered due to symmetry of the problem. There are four corner nodes A, B, C, and D as shown in Fig. 7. Three methods are compared here: (a) method with continuous boundary variables (MB = NB = 24); (b) composite boundary element method using constant and linear elements (MB=20, NB=24); and (c) boundary point element method (MB = NB = 20). The results are shown in table 2.

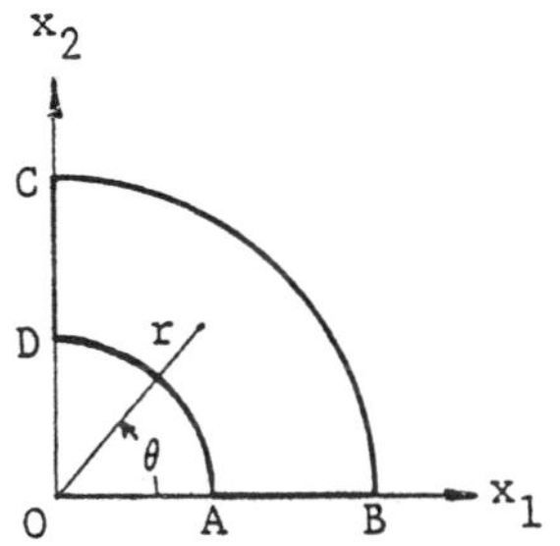

Fig.7

Table 2

Point r	θ		(a)	(b)	(c)	Exact
1	45	u	93.75	70.05	70.16	70.25
2	45	u	66.06	48.66	48.49	48.60
3	45	u	46.23	35.94	35.80	35.96
4	45	u	31.51	26.85	26.79	26.96
5	45	∂u/∂n	-10.35	-6.18	-6.26	-6.24

Thin square plate subject to uniformly distributed load

The length of each side of the plate is 2 and the prescribed traction q shown in Fig. 10a is 1. Young's modulus E = 2 and Poisson's ratio ν = 0.17. The results of displacements u_i and stresses σ_{ij} are given in Table 3 by using (a) the method with continuous boundary variables (MB = 8, NB = 9), and (b) the boundary point element method (MB = 3, NB = 4).

The following cases of prescribed displacements in using boundary point element method are also calculated to examine the validity of Eqs. (13) and (16): (1) $u_x^1 = u_x^2 = u_x^3 = 0.5$, $u_x^4 = 0$, $u_y^1 = 0$, $u_y^2 = u_y^3 = u_y^4 = -0.085$; (2) $u_x^1 = u_x^2 = u_x^3 = 0.5$, $\tau_x^4 = 0$, $u_y^1 = 0$, $u_y^2 = u_y^3 = u_y^4 = -0.085$; (3) $\tau_x^1 = 1$, $u_x^2 = u_x^3 = 0.5$, $\tau_x^4 = 0$, $\tau_y^1 = 0$, $u_y^2 = u_y^3 = -0.085$, $\tau_y^4 = 0$. The superscripts represent nodal numbers (Fig. 8b). The results of the three cases are all the same as those of row 2 in table 3. It appears that they agree with the exact solutions very well.

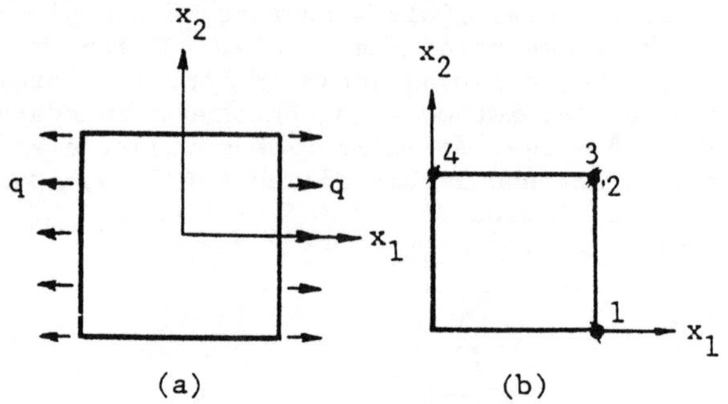

Fig.8

Table 3

Point	x	0.5	0.5	0	1	1	0
	y	0	0.5	0.5	0	1	1
u_x	(a)	0.2395	0.2679	0.0000	0.4879	0.8018	0.0000
	(b)	0.2500	0.2500	0.0000	0.5000	0.5000	0.0000
	Exact	0	0.25	0	0.5	0.5	0
u_y	(a)	-0.0000	-0.0309	-0.0160	0.0000	-0.2414	-0.0585
	(b)	0.0000	-0.0425	-0.0425	0.0000	-0.0850	-0.0850
	Exact	0	-0.0425	-0.0425	0	-0.085	-0.085
σ_x	(a)	0.9676	1.048	1.104	1.000	1.317	1.352
	(b)	1.000	1.000	1.000	1.000	1.000	1.000
	Exact	1	1	1	1	1	1
σ_y	(a)	0.0331	0.0651	0.0861	-0.2416	-0.0970	0.0000
	(b)	0.0000	-0.0000	0.0000	0.0000	0.0000	0.0000
	Exact	0	0	0	0	0	0
τ_{xy}	(a)	0.0000	0.0404	0.0000	0.0000	0.0970	0.0000
	(b)	0.0000	-0.0000	0.0000	0.0000	0.0000	0.0000
	Exact	0	0	0	0	0	0

<u>Thin square plate with circular hole subject to uniformly distributed load</u>

The plate is 4 x 4 with a hole of radius 0.5 at the center. The elastic parameters $E = 10^4$ and $\nu = 0.3$. The results of using (a) method with continuous boundary variables (MB = 16, NB = 18); and (b) boundary point element method (MB = 9, NB = 11) are compared in Table 4.

CONCLUSION

The treatment for corner node problems is essential in a boundary element analysis, especially when the analysis is applied to the engineering problems with more complicated

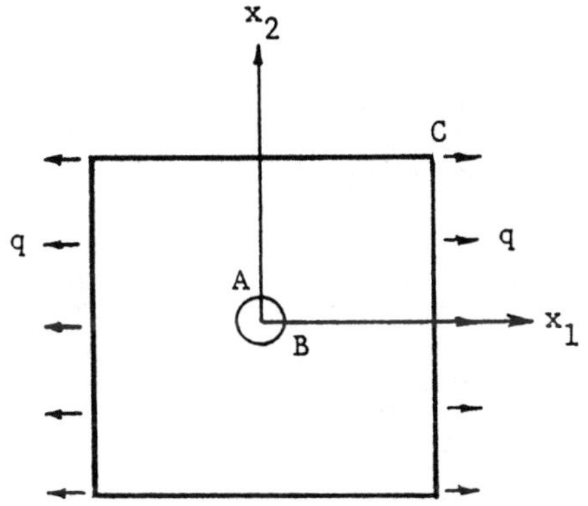

Fig.9

Table 4

Point		A	B	C
	x	0	0.5	4
	y	0.5	0	4
σ_x	(a)	2.914	0.000	1.315
	(b)	2.981	0.000	1.000
	Exact	3	0	1
σ_y	(a)	0.000	-0.691	-0.100
	(b)	0.000	-1.009	0.000
	Exact	0	-1	0
τ_{xy}	(a)	0.000	0.000	0.100
	(b)	0.000	0.000	0.000
	Exact	0	0	0

boundary conditions.

In general, the method with continuous boundary variables can be used in the situation where the difference of tractions on two sides of a corner node is not too big . In this case, the results of the tractions represent a certain "mean" effect. However it is more effective to use the boundary point element method because a satisfactory result can be achieved by using less elements. The other treatments discussed in this paper are generally better than the method with continuous variables, but still not so good as the boundary point element method.

In addition, it is shown that the proposed formula for calculating stresses at a corner node and the procedure to deal with the singularity of matrix $\underset{\sim}{A}$ are efficient and adequate. The boundary point element method can be easily implemented in a boundary element program.

REFERENCES

1. Brebbia, C.A., The Boundary Element Method for Engineers, Pentech Press, London, 1978.
2. Brebbia, C.A. and Walker, S., Boundary Element Techniques in Engineering, Butterworths, London, 1980.
3. Cruse, T.A., Application of the Boundary Integral Equation Solution Method in Solid Mechanics, International Variational Methods in Engineering, Vol.3, 1972.
4. Brebbia, C.A., Telles, J.C.F., and Wrobel, L.C, Boundary Element Techniques, Theory and Applications in Engineering, Springer Verlag, Berlin, Heidelberg, New York, Tokyo, 1984.
5. Banerjee, P.K. and Butterfield, R., Boundary Element Methods in Engineering Science, 1981
6. Hartmann, F., Computing the C-matrix in Non-smooth Boundary Point, in New Developments in Boundary Element Methods, Ed., C.A. Brebbia, pp. 367-379, 1983
7. Brebbia, C.A., Topics in Boundary Element Research, Volume 1: Basic Principles and Applications, Springer-Verlag, Berlin, Heidelberg, New York, Tokyo, 1984
8. Xu, J.M. and Brebbia, C.A., Optimum Position for the Nodes in Discontinuous Boundary Elements, 8th Int. Conf. on BEM in Eng., Tokyo, Japan, 1986
9. Brebbia, C.A.,(Ed.), Recent Advances in Boundary Element Methods, Proc. 1st Int. Conference Boundary Element Methods, Southampton University, Pentech Press, London, 1978
10. Du, Q.H. and Yao, Z.H., Some Basic Problems and Engineering Applications of Boundary Integral Equation-Boundary Element Method in Elasticity, Journal of Solid Mechanics, Vol.1, pp. 1-22, 1982
11. Yan, G., Ding, F.M., Composite Boundary Element Methods, to be published
12. Timoshenko, S.P. and Goodier, J.N., Theory of Elasticity, 3rd edition, McGraw-Hill, New York, 1970
13. Yan, G. and Ding, F.M., Basis of Boundary Element Methods, Press of Chongqing Univ., Chongqing, 1986

Accurate Integration of Singular Kernels in Boundary Methods

M. Cerrolaza*, E. Alarcón** and J. Molina**

*Instituto de Materiales y Modelos Estructurales, Universidad Central de Venezuela, PO Box 50361, Caracas, Venezuela

**Dpto. de Mecánica Estructural y Construcciones Industriales, Univ. Politécnica, Gutierrez Abascal 2, Madrid 28006, Spain

ABSTRACT

Numerical integration of singular kernels existing in the Cauchy Principal Value (CPV) sense plays a fundamental role in BEM computations since the diagonal coefficients of the influence matrix largely affect the whole solution. Moreover, in the recently developed p-adaptive BEM, the evaluation of the mathematical tools which govern the refinement process -local indicators and global estimator- needs to be carefully treated if one expects to get accurate enough results.

Thus, in this paper a fully numerical procedure for evaluating singular kernels in the CPV sense is presented and discussed. The proposed method is developed by taking into account the theoretical definition of CPV integrals and it is entirely based on standard Gauss-quadrature formulae. The gaussian sampling points are non-linearly mapped around the singularity, thus providing a great numerical accuracy. Other types of singularities, such as logarithmic and generalized CPV sense ones, can also be effectively treated without modifications of the formulae. Some numerical examples which demonstrate the accuracy of the proposed approach are included.

INTRODUCTION

Recently, it has been shown how the self-adaptive techniques can be successfully applied within the BEM context [1,2,3,4,5,6,7,8,9]. However, although the numerical results are certainly spectacular, it has been found that the poor accuracy in the numerical evaluation of both the nearly singular kernels and the singular ones (i.e., those which only exists in the Cauchy Principal Value (CPV) sense), is a problem which still remains in the BEM. This lack of accuracy especially affects the BEM p-adaptive version, since the macro-elements used to discretize the boundary are chosen as large as possible, provided they do not violate the requirements imposed by the geometry and the boundary conditions.

Furthermore, the evaluation of the residual function in the p-version, necessary for the determination of error parameters which govern the refinement process [2,10,11], is a subject that requires the greatest accuracy in the calculation of the variables involved, in order to obtain a reliable refinement criteria.

In the case of nearly-singular integrals (i.e., when the source point closely approaches the interval of integration), some special precautions must be taken. Various schemes have been proposed in the technical literature. Jun et.al. [12] subdivide the interval of integration at the regions close to source point and then perform the integration in each subdivision by means of standard Gauss-Legendre quadratures. In another publication [13], these authors employ a double-exponential formula combined with standard Gauss quadratures. These techniques, however, are expensive since they require the use of a high number of sampling points to produce acceptable results. In his interesting paper, Telles [14] presents a non-linear coordinate transformation which conveniently gathers the sampling points nearby the source point, thus giving great accuracy. This scheme is very attractive due to the simplicity of the expressions developed, and it works well enough when evaluating the state of stresses at points placed very close to the boundary.

In the case of singular kernels existing in the CPV sense (i.e., those in which the source point belongs to the interval of integration), one must proceed carefully when performing their evaluation. Some authors, such as Rank [15] and Parreira [7], have employed analytical integration over straight boundary elements. In the p-adaptive BEM version, however, such a technique proves inadvisable since it notably reduces the possibility of selecting curved macro-elements which must fit large regions of the boundary.

The numerical quadratures developed by Kutt [16,17] (finite-part quadratures) can also be used, but they need an element subdivision and are not applicable in the presence of singularities of order O(Log x). It is also possible to use the so-called "bootstrap" techniques [18], recently applied to the BEM by Giuggiani et.al. [19]. These techniques remove the singularity in the following way:

$$\int_a^b \frac{f(x)}{x-\lambda}dx = \int_a^b \frac{f(x)-f(\lambda)}{x-\lambda}dx + f(\lambda)\int_a^b \frac{dx}{x-\lambda} =$$

$$= \int_a^b \frac{f(x)-f(\lambda)}{x-\lambda}dx + f(\lambda)\,\mathrm{Log}\left|\frac{b-\lambda}{\lambda-a}\right|$$

where f(x) is a regular function in [a,b], $\lambda \in (a,b)$ and now the integral is regular, so it is possible to integrate it by means of standard Gauss-Legendre quadratures, provided that no sampling point coincides with the singular point λ. Nevertheless, these techniques have the disadvantage of not being applicable in the presence of singularities O(Log x) and $O(1/x^{\alpha})$ with $\alpha > 1$.

As is well known in the bi-dimensional elastostatics case, the fundamental solutions [20,21] $T^*(P,Q)$ and $U^*(P,Q)$ contain singularities of order O(1/x) and O(Log x) respectively. The former is hardly integrable in the CPV sense while the integration of the second must be performed by splitting the singular kernel and then applying special quadratures with logarithmic weighting functions [22]. It is clear that the aforementioned procedure implies a certain number of added numerical and computational complications.

On the other hand, the macro-elements selected to discretize the boundary geometry in the p-adaptive version [2,10,11] are chosen as large as possible and, of course, they must avoid incompatibilities with both the geometry and the boundary conditions. So, one can understand the great significance when evaluating the singular kernels in order to obtain a reasonable enough accuracy. In addition, the calculation of the residual function requires the "a-posteriori" computation of the variables inside the boundary elements, upon which the numerical integration must be carefully performed.

Thus, the aim of this paper is to develop a fully numerical procedure to evaluate both the singular kernels existing in the CPV sense and those containing singularities O(Log x) which normally arise in boundary methods.

THE BI-CUBIC TRANSFORMATION

Let the integral

$$I = \int_a^b \frac{x - x_s}{|x - x_s|^{\alpha}} \cdot w(x)\, f(x)\, dx \qquad (1)$$

$$x_s \in (a,b)\ ; \quad \alpha \in \{\mathbb{R}^+\},\ \alpha >= 2$$

where $\int$ denotes "finite part integral", f(x) is any regular function differentiable enough in [a,b], w(x) is an arbitrary weighting function also defined in [a,b] and x_s is the singular point. Assuming that $f(x) \neq 0$, and if $\alpha = 2$, then integral (1) exists in the CPV sense. In the case of $\alpha > 2$, f(x) needs to meet some additional conditions [16,23] in order to be considered a "generalized Principal Value integral". In the following discussion, we will restrict ourselves to the case $\alpha = 2$, although the proposed algorithm is also applicable even when $\alpha > 2$. Thus, integral (1) could be rewritten as [16]:

$$I = \lim_{\varepsilon \to 0^+} \left[\int_a^{x_s - \varepsilon} g(x)\, dx + \int_{x_s + \varepsilon}^{b} g(x)\, dx \right] \tag{2}$$

where the kernel g(x) now contains the singularity of order α

$$g(x) = \frac{x - x_s}{|x - x_s|^{\alpha}} \; w(x)\, f(x) \tag{3}$$

Integral (2) must be mapped into a domain $\eta \in [-1,1]$ in order to properly perform its numerical evaluation by means of standard Gauss quadratures. So, expression (2) becomes:

$$I = \lim_{\varepsilon \to 0^+} \left[\int_{-1}^{\eta_s - \varepsilon} G(\eta)\, d\eta + \int_{\eta_s + \varepsilon}^{1} G(\eta)\, d\eta \right] \tag{4}$$

where now η_s is the image of the singularity x_s. The central idea of the numerical procedure proposed here is based on the following two requirements :

a) it is advisable to gather the sampling points nearby the singularity in order to improve the numerical representation of the high gradients and

b) it is necessary to obtain a sampling point distribution which essentially respects the demands of eq. (4), i.e., the quantity ε must be the same at both sides of singular point η_s.

A non-linear coordinate transformation which satisfies requirement a) has been proposed in ref. [14]. Such a transformation works well in cases where the singular point is centered on the interval of integration ($\eta_s = 0$) by reason of the symmetry of the standard Gauss-Legendre quadrature. However, when the singularity is not centered at the interval of integration ($\eta_s \neq 0$) the aforementioned transformation does not satisfy requirement b), thus leading to an inaccurate results.

For the sake of clarity and with no loss of generality, consider the simplest case, i.e., f(x)=1 and w(x)=1. Thus, expression (4) could be rewritten as

$$I = \lim_{\varepsilon \to 0^+} \left[\int_{-1}^{\eta_s - \varepsilon_a} \frac{\eta - \eta_s}{|\eta - \eta_s|^{\alpha}} d\eta + \int_{\eta_s + \varepsilon_b}^{1} \frac{\eta - \eta_s}{|\eta - \eta_s|^{\alpha}} d\eta \right] \quad (5)$$

It is clearly understood that if requirement b) is not properly satisfied, the undesirable condition $|\varepsilon_a| \neq |\varepsilon_b|$ will arise and unreliable results will be obtained.

In the bicubic transformation proposed, the position of the sampling points is modified according to two cubic functions (hence the name BCT) which are defined over both sides of the singularity. Thus, two polynomials are introduced:

$$\eta_a(\xi) = A_a\xi^3 + B_a\xi^2 + C_a\xi + D_a \; ; \qquad \eta_b(\xi) = A_b\xi^3 + B_b\xi^2 + C_b\xi + D_b \quad (6)$$

where the suffix "a" now denotes the previous region (to the singularity) of the interval of integration, whereas suffix "b" denotes the latter region. The following boundary conditions are found to be necessary:

$$\eta_a(\xi = -1) = -1 \; ; \qquad \eta_a(\xi = 1) = \eta_s \quad (7)$$

$$\eta_a(\xi_l) = \eta_s - \varepsilon \quad (8)$$

$$\left. \frac{d\eta_a}{d\xi} \right|_{\xi = 1} = 0 \quad (9)$$

$$\eta_b(\xi = -1) = \eta_s \; ; \qquad \eta_b(\xi = 1) = 1 \quad (10)$$

$$\eta_b(\xi_f) = \eta_s + \varepsilon \quad (11)$$

$$\left. \frac{d\eta_b}{d\xi} \right|_{\xi = -1} = 0 \quad (12)$$

As can be seen, restraints (7) and (10) impose, respectively, the new integration bounds for both the previous and the latter region. Restriction (8) forces the image of the last point (ξ_l) in the gaussian quadrature used in the previous part (see fig. 1) to be placed at a distance ε from the singular point. In the same way, the image of the first point (ξ_f) in the gaussian quadrature for the latter part is forced to be placed at a distance ε from the singular point. As a consequence, the nearly-infinite values which normally occur when evaluating the singular kernel at points $\eta_s+\varepsilon$ and $\eta_s-\varepsilon$ will display the same numerical value and opposite sign, thus being conveniently cancelled.

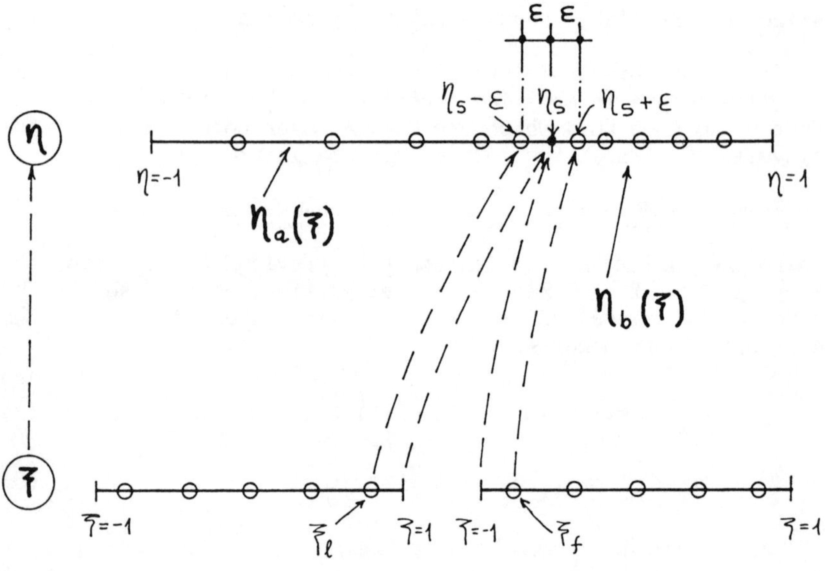

Fig 1. New distribution of sampling points produced by the BCT

Restraints (9) and (12) smooth out the singularity near the singular point and, therefore, even better accuracy is reached. Figure 1 shows the effect that transformations (6) produce over a standard Gauss-Legendre quadrature distribution of sampling points.

The application of the previous boundary conditions to eq. (6) yields the following polynomial coefficients :

$$A_a = \frac{1+\eta_s - \varepsilon + \frac{1+\eta_s}{4}(\xi_l^2-1) - \frac{1+\eta_s}{2}(\xi_l+1)}{\xi_l^3 - \xi_l^2 - \xi_l + 1}$$

$$B_a = \frac{4A_a+1+\eta_s}{4} \quad ; \quad C_a = \frac{1+\eta_s}{2} - A_a \quad ; \quad D_a = -1+A_a-B_a+C_a \tag{13}$$

$$A_b = \frac{\varepsilon - \frac{1-\eta_s}{4}(\xi_f^2-1) - \frac{1-\eta_s}{2}(\xi_f+1)}{\xi_f^3 + \xi_f^2 - \xi_f - 1}$$

$$B_b = \frac{4A_b+1-\eta_s}{4} \quad ; \quad C_b = \frac{1-\eta_s}{2} - A_b \quad ; \quad D_b = 1-A_b-B_b-C_b \tag{14}$$

Thus, if we have an integral of the form $\int_{-1}^{1} f(\eta)d\eta$ containing a singular point $\eta_s \in (-1,1)$, its numerical evaluation by means of the BC transformation will be as follows

$$I = \int_{-1}^{1} \Big[f[\eta_a(\xi)]\, J_a(\xi) + f[\eta_b(\xi)]\, J_b(\xi) \Big]\, d\xi =$$

$$= \sum_{k=1}^{NGP} \Big[f[\eta_a(\xi_k)]\, J_a(\xi_k) + f[\eta_b(\xi_k)]\, J_b(\xi_k) \Big]\, w_k \tag{15}$$

where the ξ_k and the w_k are the abscissas and the weighting factors of the gaussian quadrature of order NGP. The use of the same order of quadrature in both the previous and the latter regions of the interval of integration is implicit in eq. (15). This particularity, even though it is not binding since it is perfectly possible to use different orders, has been shown to be highly efficient and simplifies the computational details. Also investigated and developed were other non-linear transformations of higher order (up to 5). Even though acceptable results were obtained, a certain loss of accuracy is observed due to the strong gradients displayed by these functions in the neighbourhood of the singularity, thus leading to a certain shifting of the new weighting factors from the desired nearly-zero value [10].

On the other hand, our experience has shown that a numerical value of the parameter ε approximately equal to one thousandth the size of the interval of integration gives results accurate enough for practical applications in p-adaptive BIEM.

The source point coincides with one end of the element.

As it is well known, inaccurate results are obtained if two boundary elements of very different sizes are defined meeting a common source point and if no special precautions are taken regarding the integration procedure.

In figure 2.a it can be noted that if we do not appropriately satisfy the condition $|d_k| = |d_j|$, a very strong numerical contribution will normally exist at source point P which will lead to unstable results. However, the BC transformation can also be used in such a case, since it is represented by a CPV integral extended over two boundary elements. Thus, when dealing with these situations and depending upon the position of the singular point η_s, the following applies:

if $\eta_s = -1$ then use $\eta_b(\xi)$; if $\eta_s = 1$ then use $\eta_a(\xi)$

In order to get the same d-values (see fig. 2.b) around the source point P, the following procedure is suggested :

1. Compute an ε_k related to the smallest element meeting the source point (usually $\varepsilon_k = L_k/1000$)
2. Compute another ε_j related to the largest element meeting the source point, in such a way to ensure that $|d_k| = |d_j|$ in the real space.

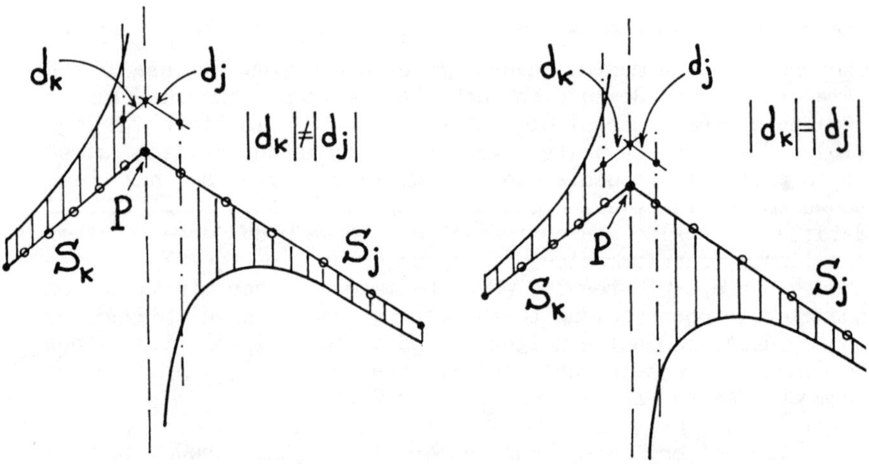

Fig. 2. Elements of different size meeting a source point P.

These operations are rather trivial on a computer and can also be quickly added to any existing BEM code. In the same way, expressions (6) can be easily embodied into operative numerical integration routines. The increment in computer time is negligible when compared with traditional schemes based upon standard Gauss-Legendre quadratures.

The BC transformation could also be used in integration contexts other than BEM by simply adjusting the parameter ε and/or by using proper orders of integration in gaussian quadratures.

NUMERICAL EXAMPLES

Some numerical examples, which include singularities of order O(Log x), $O(1/x^{\alpha})$ and of practical application in p-adaptive BEM are included herein in order to show the great accuracy obtained by using the BC transformation.

a) $$I = \int_{-1}^{1} \text{Log} \mid \eta - 0.8 \mid d\eta \qquad (16)$$

This simple example contains a singularity O(Log x) at η = 0.8. Table I collects the evolution of both the numerical results as well as the errors produced (in absolute value) when using standard Gauss-Legendre quadrature, the method of Ref. [14] and the BC transformation. The results are given as a function of the gaussian quadrature order NGP.

NGP	Gauss	Gauss/ Analyt.	Ref 14	Ref 14/ Analyt.	BCT	BCT/ Analyt.
4	-1.2165	0.9625	-1.1428	0.9042	-1.4275	1.1295
6	-1.0802	0.8547	-1.2605	0.9973	-1.2757	1.0094
8	-1.7911	1.4171	-1.2762	1.0097	-1.2646	1.0006
10	-1.1442	0.9053	-1.2675	1.0028	-1.2637	0.9999
12	-1.2518	0.9904	-1.2599	0.9968	-1.2638	0.9999
14	-1.2283	0.9719	-1.2623	0.9988	-1.2638	1.0000
16	-1.1964	0.9466	-1.2652	1.0010	-1.2639	1.0000

Analytical value = -1.2638715857 E+000

Table I. Numerical evaluation of singularity O(Log x)

As can be observed, satisfactory results are obtained by employing only six integration points (3 in the previous region to the singularity and 3 in the latter one).

b) $$I = \int_{-1}^{1} \frac{\text{SGN}(\eta-0.3)}{|\eta-0.3|} \, d\eta \qquad (17)$$

where SGN means "sign of". As in the previous case, integral (17) was evaluated by using standard Gauss-Legendre quadrature, method of Ref. [14] and the BC transformation. Table II contains both the numerical results as well as the error produced (in absolute value), as functions of the order of gaussian quadrature NGP.

NGP	Gauss	Gauss/ Analyt.	Ref 14	Ref 14/ Analyt.	BCT	BCT/ Analyt.
4	15.6130	25.221	4.1241	6.662	-8.5274	13.775
6	-7.7365	12.498	6.8693	11.097	-1.4164	2.288
8	-2.5327	4.091	10.7110	17.302	-0.7396	1.195
10	-0.4259	0.688	17.1290	27.670	-0.6517	1.053
12	1.8709	3.022	31.8650	51.475	-0.6308	1.019
14	9.8459	15.905	122.270	197.509	-0.6240	1.008
16	-10.1820	16.448	-73.087	118.065	-0.6214	1.004

Analytical value = -6.1903920841 E-001

Table II. Numerical evaluation of singularity O(1/x)

Once again, the stability and accuracy of the BC transformation can be observed. As can be seen, method in Ref. [14] that worked very well in the previous example, produces completely unacceptable results here. That kind of behaviour was the main reason to try to improve it, as discussed above, by imposing a total symmetry of the two integration points neighbour to the singular one. In this way, we enforce a numerical interpretation of the definition of the singularity in the CPV sense.

c)
$$I = \int_{-1}^{1} \frac{\mathrm{SGN}(\eta-0.2)}{|\eta-0.2|^{1.2}} \, \mathrm{EXP}(\eta)\, d\eta \qquad (18)$$

The integral shown above displays a singularity of order $\alpha = 1.2$ at $\eta = 0.2$ and it is included herein in order to assess the applicability of the proposed numerical procedure even when $\alpha > 1$. Table III contains both the numerical results and the errors (in absolute value) obtained by using the aforementioned methods, as functions of the order of gaussian quadrature NGP.

As expected, numerical results show the accuracy of the BC transformation even in the presence of singularities $O(1/x^{\alpha})$ with $\alpha > 1$. Relative errors of approximately 1 % were obtained by employing only 10 Gauss integration points for the whole interval.

d) Displacement calculation on a square plate.

Figure 3 shows a square plate subjected to a parabolic variation of the boundary tractions. The dimensions of the domain are 100x100 cms^2, its thickness is t=1 cm., while the

NGP	Gauss	Gauss/ Analyt.	Ref 14	Ref 14/ Analyt.	BCT	BCT/ Analyt.
2	4.9770	2.034	6.6120	2.703	-114.4700	46.792
4	9.9416	4.064	12.5180	5.117	-3.6379	1.487
6	30.6170	12.515	21.5240	8.798	1.8527	0.757
8	-57.7600	23.610	34.4090	14.065	2.3340	0.954
10	-9.3260	3.812	53.1210	21.714	2.4161	0.988
12	-2.5264	1.033	81.5850	33.349	2.4381	0.997
14	84.6270	0.346	128.3000	52.444	2.4463	1.000

Finite-Part (Kutt, 1975) = +2.4464143506 E+000

Table III. Numerical evaluation of singularity $O(1/x^{1.2})$

Young's modulus was taken as $E=2.1x10^6$ Kgf/cm^2 and Poisson's coefficient as $\nu=0$. This problem was analysed through the p-adaptive BEM version [4,10]. Its aim is to evaluate the displacement tensor at boundary points inside the elements in order to further calculate the residual function.

NGP		Gauss	Ref [14]	BCT	T1/T3	T2/T3
6	u_1	1.0874	2.8576	0.9579	1.10	1.04
	u_2	0.6503	6.4416	1.2923		
8	u_1	1.0874	37.8201	0.9927	1.06	1.00
	u_2	0.6503	146.2905	1.0315		
10	u_1	1.0874	1.0982	0.9978	1.06	1.00
	u_2	0.6503	9.3885	1.0106		
12	u_1	1.0874	0.1223	0.9992	1.07	1.00
	u_2	0.6503	4.5103	1.0051		
14	u_1	1.0874	0.5905	0.9997	1.07	1.00
	u_2	0.6503	2.6389	1.0031		

$u_j = u_j^{comp} / u_j^{theory}$

T1 = Time for Gauss-Legendre
T2 = Time for method of Ref [14]
T3 = Time for BC transformation

NGP = Number of Gauss points used on element 1.

Table IV. Normalized displacements on square plate

Table IV contains the displacement normalized with regard to the theoretical value at point P on element 1, as a function of the gaussian quadrature order. Also shown is a comparison of the computer time employed by the three methods. The values obtained with the BC transformation by using 8 sampling points on element 1 and 6 sampling points on all the others (26 points for the whole boundary) are accurate enough for practical purposes in the p-adaptive BEM. The computer time is sligthtly lesser than that which corresponds to standard gaussian quadrature, since in the latter scheme it is necessary

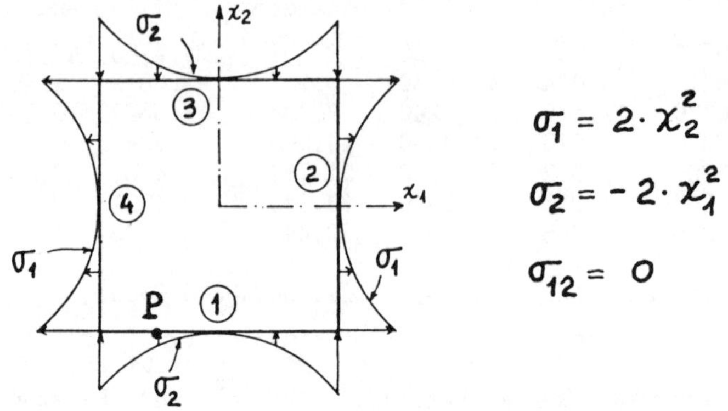

Fig. 3. Square plate: boundary mesh and tractions.

to split the singular kernel depending upon the type of the singularity, whereas in the BC transformation it is not.

CONCLUDING REMARKS

A new numerical procedure for the evaluation of CPV integrals and those which contain other type of singularities has been presented.

The mapping proposed herein gathers the sampling points in the neighbourhood of the singularity in order to catch the high gradients around it as in Ref. [14], but uses two different third order mappings around the singularity in order to properly enforce the total symmetry of the two integration points neighbour to the singularity in the real space. As shown by the examples, that symmetry is essential for the success of the procedure.

Numerical examples have shown that the proposed method is stable and gives results accurate enough for practical applications in p-adaptive BEM. The formulas are rather simple. They make use of standard Gauss-Legendre quadratures and can also be quickly embodied into already existing integration routines.

On the other hand, it is not necessary to split the singular kernels which contain singularities of different orders, which notably simplifies the usual numerical procedures employed in BEM. Moreover, the formulas are capable of accurately integrating those singular functions displaying singularities of order $O(1/x^{\alpha})$, with $\alpha > 1$.

ACKNOWLEDGEMENTS

The authors wish to express their thanks to the support provided by the Consejo de Desarrollo Cientifico y Humanistico (CDCH, Venezuela) and the Instituto de Cooperacion Iberoamericana (ICI, España).

REFERENCES

[1] Alarcon, E. and Reverter, A., "P-adaptive Boundary Elements", Int. J. Num. Met. Eng., 23, 801-829 (1986)

[2] Alarcon, E., Reverter, A. and Molina, J., "Hierarchical boundary elements",Comp.& Struct., 20(1-3), 151-156,1986

[3] Cerrolaza, M. and Alarcon, E., "p-adaptive boundary elements for three dimensional potential problems", Comm. Applied Num. Methods, 3, 335-345 (1987)

[4] Cerrolaza, M. and Alarcon, E., "Elastostatics p-adaptive boundary elements for micros", Soft. Eng. Workstations, 4,18-26 (1988)

[5] Cerrolaza, M. and Alarcon, E., "The p-adaptive BEM approach in 2-D elasticity", Microcomp. in Civil Eng., to be published (1989)

[6] Cerrolaza, M. and Alarcon, E., "Further applications of p-adaptive BEM", Proc. X Int. Conf. on BEM, (Brebbia et. al. eds.), Springer-Verlag, Southampton (1988)

[7] Parreira, P., "Self-adaptive p-hierarchical boundary elements in elastostatics", Proc. IX Int. Conf. on BEM (Brebbia et.al. Ed), Springer-Verlag, Stuttgart (1987)

[8] Rank, E., A posteriori fehlerabschatzungen und adaptive Netzverfeinerung fur Finite Element und Randintegralelement-Methoden, Ph.D. Thesis, Univ. of Munich (1985)

[9] Rencis, J.J. and Mullen, R.L., "Solution of elasticity problems by a self-adaptive mesh refinement technique for boundary element computation", Int. J. Num. Met. Eng., 23, 1509-1527 (1986)

[10] Cerrolaza, M., p-adaptive boundary elements: development and applications in Potential Theory and elastostatics, Doctoral Thesis (in spanish), Madrid (1988)

[11] Gomez, M., Cerrolaza, M. and Alarcon, E., "Adaptive refinements in boundary element methods", Proc. IX Int. Conf. on BEM, Springer-Verlag, Stuttgart (1987)

[12] Jun, L., Beer, G. and Meek, J.L., "Efficient evaluation of integrals of order 1/r, 1/r2, 1/r3 using Gauss quadrature", Engineering Anal., 2, 118-123 (1985)

[13] Jun, L., Beer, G. and Meek, J.L., "The application of double-exponential formulas in the boundary element method", Proc. VII Int. Conf. on BEM (Brebbia and Maier Eds.), Springer-Verlag, Berlin (1985)

[14] Telles, J.F.C., "A self adaptive coordinate transformation for the efficient evaluation of general boundary element integrals", Int. J. Num. Met. Eng., 24,

937-959 (1987)

[15] Rank, E., "Adaptive Boundary element methods", Proc. IX. Int. Conf. on BEM, (Brebbia et.al. Eds.), Springer-Verlag, Stuttgart (1987)

[16] Kutt, H.R., "The numerical evaluation of Principal Value integrals by finite-part integration", Num. Math., 24, 205-210 (1974)

[17] Kutt, H.R., "Quadrature formulas for finite-part integrals", Report Wisk 178, The Nat. Institute for Math. Sci., Pretoria (1975)

[18] Rabinowitz, P., "Some practical aspects in the numerical evaluation of Cauchy Principal Value integrals", Seminar on Numerical Analysis (Michavila Ed.), Madrid (1987)

[19] Giuggiani, M. and Casalini, P., "Direct computation of Cauchy Principal Value integrals in advanced boundary elements", Int. J. Num. Met. Eng., 24, 1711-1720 (1987)

[20] Brebbia, C.A., Telles, J.C. and Wrobel, L.C., Boundary Element Techniques: Theory and Applications in Engineering, Springer-Verlagm Berlin (1984)

[21] Cruse, T.A., "Mathematical foundations of the BIEM in solid mechanics", Report AFOSR-TR-77-1002, Pratt & Whittney Aircraft Group, USA (1977)

[22] Berthold-Zaborowsky, H., "Calcul des integrales de la forme $\int_0^1 \mathrm{Log}(x)\, f(x)$ ", Techniques de Calcul Numerique, H. Mineur, Dunod, Paris (1966)

[23] Paget, D.F. and Elliot, D., "An algorithm for the numerical evaluation of certain Cauchy Principal Value integrals", Numerische Mathematik., 19, 373-385 (1972)

SECTION 4 - COMPUTATIONAL ASPECTS

Numerical Experiments with a Boundary Element Technique for Corners

L.J. Gray

Mathematical Sciences Section, Oak Ridge National Laboratory, Oak Ridge, TN 37831, USA

ABSTRACT

A method for handling regions with corners or discontinuous boundary data has been recently proposed by Mitra and Ingber (*Boundary Elements IX, pp. 519-532*). Using double nodes to represent the corner and an external collocation point to generate the necessary additional equation, this paper reported (for the most part successful) test results on a variety of two dimensional problems. In the present paper, an exactly solvable Laplace equation problem, posed in a region with reentrant and non-reentrant edges, has been employed to study the behavior of this algorithm in three dimensions. As boundary surface edges give rise to multiple corner points, this problem provides a challenging test for the proposed alorithm. Experiments using several sets of external collocation points indicate that this technique is not a reliable procedure in three dimensions, yielding answers that are sensitive to the location of external nodes.

INTRODUCTION

It is well known that normal derivative values can become singular at a corner or edge of the boundary surface.[1] However, even if the solution is *not* divergent as the corner is approached (i.e., the one sided limits exist), accurate boundary element results are still difficult to obtain in the vicinity of the non-smooth boundary.[2] As this type of problem, involving surface discontinuities either in the form of geometry or discontinuous boundary data, often occur in engineering applications,[3,4] this is a serious deficiency. A complete review of efforts to define a boundary element corner procedure, and a proposed new technique, can be found in the recent paper of Mitra and Ingber.[3]

For ease of discussion, it will be convenient to focus on the simplest possible example, namely the Laplace equation $\nabla^2 V = 0$. The standard integral equation formulation is given by[5]

$$\psi(X_0)V(X_0) + \int_{\partial R} V(X)\nabla G(X,X_0)\bullet\vec{n}\,dS = \int_{\partial R} G(X,X_0)\nabla V(X)\bullet\vec{n}\,dS. \qquad (1)$$

where $\psi(X_0)$ is the interior solid angle at X_0 and $G(X,X_0)$ is the Green's function or fundamental solution. Note that the normal derivative $\nabla V(X)\bullet\vec{n}$ enters directly into the equation. In general, the flux will be discontinuous passing through a corner, and undefined precisely at the corner (where a collocation point is usually placed); in essence, this explains why a straightforward boundary element approach fails when confronted by a nonsmooth boundary surface for which the flux is the unknown quantity. The method described by Mitra and Ingber[3] recognizes the presence of the flux discontinuity and utilizes two nodes to represent a (two dimensional) corner. Clearly, Eq. (1) cannot be employed for both of these nodes (identical equations are obtained). These authors construct the additional equation by selecting an external collocation point X_0 for use in Eq. (1); if X_0 is external to the region, $\psi(X_0) = 0$, and no new unknowns are introduced by employing this equation.

Results for a variety of two dimensional problems were presented in reference 3. Although some difficulty was seen in one particular example (see pg. 526, Table 1.), good accuracy and low condition numbers were generally reported. In three dimensions however, the problem is somewhat nastier: the isolated corner point in two dimensions becomes a curve (edge) of corner points in three dimensions. It is therefore of interest to investigate the ability of this method to deal with this more difficult situation. In this paper, an exactly solvable problem with reentrant and non-reentrant edges will be examined.

COMPUTATIONAL EXPERIMENTS

Model

The region in which $\nabla^2 V = 0$ will be solved consists of that portion of the interior of a square box which is exterior to four cylinders symmetrically placed near the corners of the box. A top view of this configuration is shown in figure 1. The origin of the coordinate system is the top ($z = 0$) center of the box. Dirichlet (known potential V) boundary data will be employed, namely

$$V(x, y, z) = z^2 - \frac{x^2 + y^2}{2}. \tag{2}$$

As this function satisfies Laplace's equation, this is also the solution in the interior of the region, and the boundary flux is given by

$$\frac{\partial V}{\partial N} = -xN_1 - yN_2 + 2zN_3 \tag{3}$$

where $N = (N_1, N_2, N_3)$ is the surface normal.

The geometry and boundary conditions are symmetric with respect to x and y and with respect to the coordinate planes. Hence, it is only necessary to use a one-eighth section of the box formed by the $y = 0$ and $x = y$ symmetry planes. Figure 2 shows this reduced computational model. In this figure, the $z = 0$ plane is at the top, and the positive z axis extends down towards the floor ($z = 18$); the back wall is the $x = 18$ plane, and the cylinder is cut in half by the $x = y$ symmetry plane. Either from Eq. (3) or symmetry considerations, it follows that the flux on the three planes $y = 0$, $x = y$ and $z = 0$ is equal to zero; this information has been incorporated into the Green's function, so these surfaces are not needed in the discretization.[5,6]

Note that the bottom part of the cylinder ($z = 13.5$) is capped off by a disk. The intersection of the cylinder with this cap forms the (curved) reentrant edge in the problem; the intersection of the back wall and the floor supplies the (straight) non-reentrant

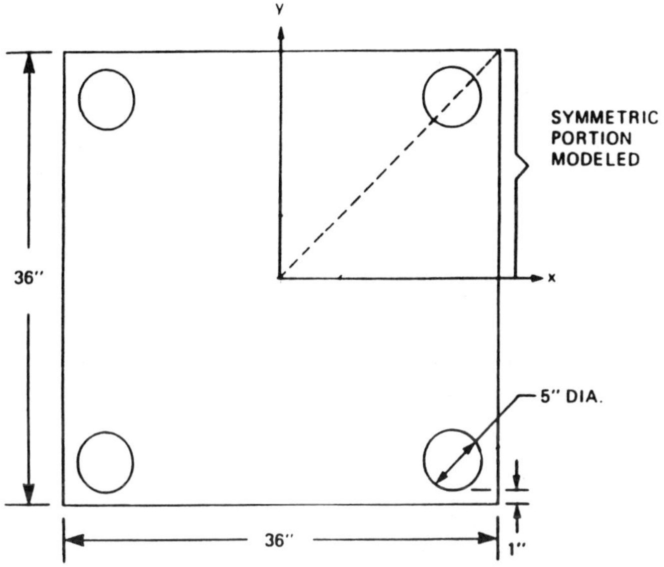

Figure 1. A top view of the problem geometry, showing the four cylinders located at the corners. For the computations, the indicated 45° symmetric section is used.

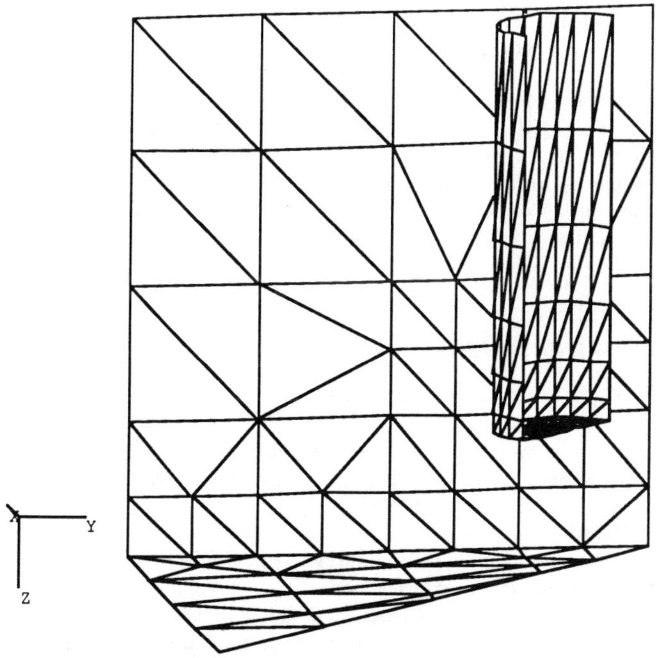

Figure 2. The computational model, showing the reentrant edge (joining the cylinder and base) and the non-reentrant edge (joining the floor and side wall).

edge. For the model in figure 2, the curved geometry along the cylinder edge was defined by 13 points whereas only 9 points were used at the junction between the wall and floor. As described in reference 3, the discretization employs double nodes at these edges, for a total of 26 and 18 nodes, respectively. Although the artificially contrived boundary conditions make this a toy problem, this configuration is nevertheless not far removed from an actual application. Figure 1 is basically an electroplating cell without a cathode, the cylinders being anodes.[6] In electroplating problems, the flux is the unknown quantity on the anode surface, and anodes with sharp edges such as the capped cylinder have been employed in practical work.

The calculations employed linear interpolation over the planar triangular elements. The only nonstandard aspect of the approximations was the use of a 20 point Gaussian quadrature to generate the external node equations. The integrations are carried out in polar coordinates (r, θ); as the r integration is performed analytically, the numerical quadrature is only for the θ direction, and a 6 point scheme is generally sufficient for most calculations. This high order scheme was invoked because it turned out to be essential to choose the external points close to the boundary surface.

Exterior Points

To complete the description of the experiments, it is necessary to indicate the location of the external collocation points. The choice of these exterior points cannot be completely arbitrary. Clearly, it is necessary to keep a "reasonable" separation between the exterior nodes, as two nodes close together produce nearly identical equations (note that there are no singular integral terms to create large diagonal matrix elements). Another constraint is that nodes cannot be placed too close to the boundary, as difficulties would arise in the accurate computation of nearly singular integrals. On the other hand, nodes placed far away generate equations which are out of scale with the rest of the matrix (all the elements being very small), leading to an ill-conditioned system. Although very imprecise, these rules provide some guidance in the choice of external points.

For a node (x, y, z) along the floor/wall, a corresponding node was placed directly below at $(x, y, z + z_f)$, $z_f > 0$. On the cylinder a small alteration of the z component does not work, as one would obtain either an interior point or a boundary point. A relatively simple algorithm for generating exterior points near the cylinder cap is to move the point up by adding z_c, $z_c < 0$, to the z component, while at the same time shrinking the radius of the cylinder. The equation of the cylinder is

$$(x - 14.5)^2 + (y - 14.5)^2 = (2.25)^2; \qquad (4)$$

an external node corresponding to (x, y, z) was placed at $(\hat{x}, \hat{y}, z + z_c)$, where

$$(\hat{x} - 14.5)^2 + (\hat{y} - 14.5)^2 = (2.25 s_c)^2. \qquad (5)$$

The vertical shifts z_f and z_c and the scaling factor s_c $(0 < s_c < 1)$ provide convenient adjustable parameters for studying the performance of the corner algorithm.

Numerical Results

The two calculated flux values at each surface edge are shown in figures 3-6 for five of the more successful sets of external nodes. (Away from the edges the computed and exact solutions agreed in these cases, and thus it suffices to examine this region.) The x-axis in these plots is just a consecutive node numbering along the edge: for the cylinder, the numbering starts at the node closest to the origin and proceeds around towards the corner of the box; for the floor, the numbering proceeds from $y = 0$ to the corner $y = 18$. The parameter values defining the exterior node locations for each calculation, together with an estimate K of the condition number of the linear system, are given in table 1.

Table 1. **External Node Parameters**

Run	z_f	z_c	s_c	K
1	0.3937	-0.3937	0.900	12,455
2	0.3937	-0.3937	0.950	2,973
3	0.1868	-0.3937	0.970	2,604
4	0.1181	-0.3937	0.990	2,611
5	0.1181	-0.5705	0.950	21,607

As might be expected, the results at the junction between the wall and the floor depend primarily on the value of z_f. Thus, runs 1 and 2 were virtually identical at this edge, as were runs 4 and 5. Consequently, the results for runs 1 (solid), 3 (dashed), and 4 (dotted) are plotted in figures 3 and 4. The exact values are constant, namely 36 for the floor and -18 on the side wall. In figure 5, the normal derivative on the cylinder cap is plotted using the same scheme as the previous figures, with the addition of run 2 (dash dot) and run 5 (asterisk); the exact value is -26. In figure 6, the solid line is now the (nonconstant) exact solution, and the results of run 1 are plotted by dots.

Figure 3. Calculated normal derivative -- Floor

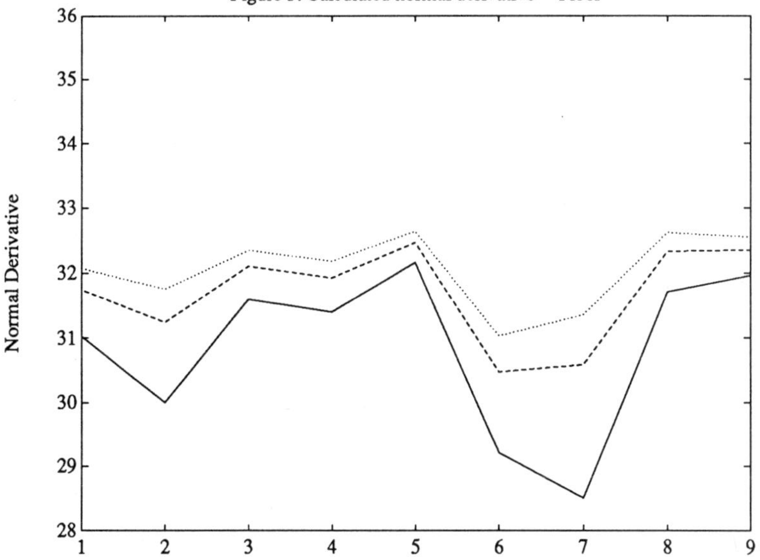

From figures 3-6 it is obvious that, although the basic discontinuity of the flux at each corner is recognized, the numerical values are not very accurate. Despite the inherent geometry error in the discretization of the curved surface, the results for the cylinder are noticeably better than those for the straight edge along the floor. However, the solution for the cylinder is much more sensitive to the choice of external points. The higher accuracy is probably due to the smaller discontinuity of the normal derivative

Figure 4. Calculated normal derivative -- Wall

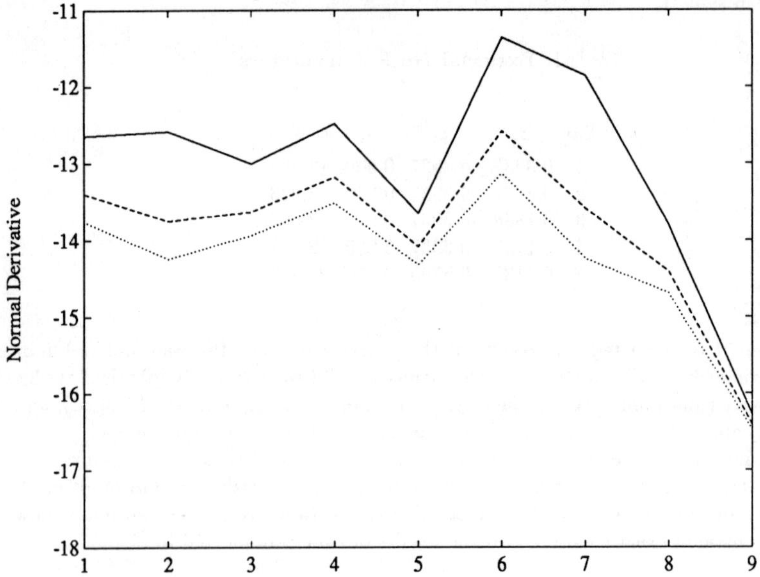

Figure 5. Calculated normal derivative -- Cap

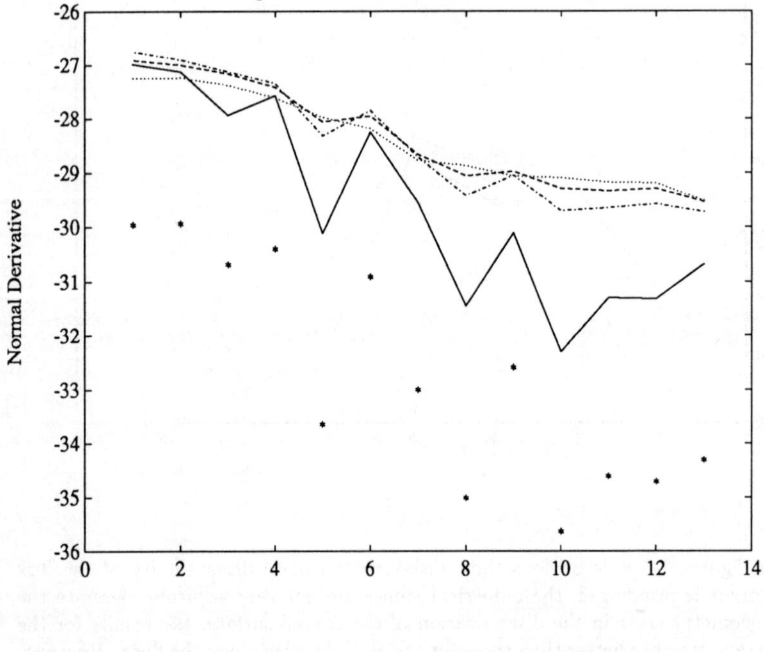

at this edge. Note that the results on the cap (figure 5) get progressively worse as the jump across the edge gets larger.

Figure 6. Calculated normal derivative -- Cylinder

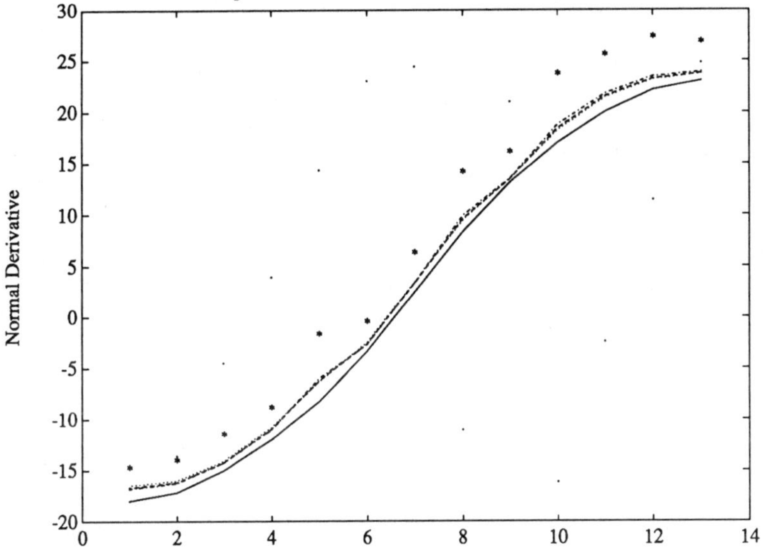

Further experimentation with the parameters did not produce any improvements. Smaller scaling of the cylinder ($s_c = .8, .7$) yielded totally incorrect results, and condition numbers on the order of 10^6. Presumably the sharp rise in the condition of the matrix is due to packing the external points closer together inside the cylinder. A similar deterioration of the results was seen by setting $z_c = -0.1868$ in run 1 (all other parameters the same), a possible indication that inaccurate evaluation of nearly singular integrals was becoming a factor. The results of run number 5 indicate that moving the cylinder nodes in the opposite direction, namely further away from the cap, also has a negative effect. This was confirmed by a test in which the external nodes for the cylinder were moved completely outside of the box (the external node for the edge node (x, y, z) was placed at $(x, y, -0.3937)$); once again, the answers were totally wrong, with a condition number of 10^{12}. From figures 3 and 4 it is clear moving the external nodes closer to the floor is beneficial; however, the solution seems to have "converged" at this point, as smaller values of z_f (as low as 0.0394) did not produce any noticeable change.

DISCUSSION

Although the selection of external node locations seemed to be a natural one for this problem, the accuracy of the solution near the boundary edges in any of the (limited number of) trial runs was not very good. However, the (reasonable) answers that were obtained are significantly better than a straightforward boundary element solution, which quite naturally tends to average the two limiting flux values at the corner. Nevertheless, from the point of view of the reliability of the algorithm, the most distressing aspect of these calculations is the observed sensitivity to the choice of external nodes.

For this problem there appears to be a very narrow range of the parameters defining the external nodes which yield reasonable results. While it is conceivable that more sophisticated functional approximations (i.e., higher order elements) or numerical integration schemes would enhance the accuracy, it seems highly unlikely that these improvements would radically alter the strong dependence upon the node locations.

Whether or not there exists a suitable (i.e., produces correct answers) choice of external points for this problem is immaterial. The sensitivity of the solution indicates that it would be very difficult to produce a generally applicable method for siting the nodes appropriately. Lacking this knowledge, any calculation using this method is subject to suspicion, seriously hindering the practical application of the algorithm.

The test results indicate that, for this particular problem, it is necessary to place the nodes close to the boundary edge (Mitra and Ingber report having to do this for one of their test cases); in addition to concerns about accurate evaluation of nearly singular integrals, this forces the external nodes (for a three dimensional problem) to be close together, possibly aggravating the conditioning of the linear system of equations. (A condition number on the order of 10^4 (see Table 1) is, in general, not a problem. However, as the boundary element approximations introduce errors in the matrix elements and the right hand side vector, this is a concern. Note that significantly better results on the cylinder were obtained when the condition number was an order of magnitude lower).

The above speculation about the causes of the failure of the method should not overlook another possibility, namely that, despite extensive checking, a bug remains in the computer code. In this regard, it should be remarked that the program is a minor reworking of a Laplace solver that has been working successfully for several years. The major new component, a subroutine for generating the external collocation point equations, required only slight modification of existing code. In addition, these new equations were checked by the simple procedure of putting in the known flux solution and checking that the calculated right hand side of the equation was approximately reproduced. This was indeed the case in all calculations. Moreover, as indicated above, away from the edges the answers were very reasonable. Thus, while it cannot be guaranteed that the program is bug free, the evidence indicates that it is performing as designed.

ACKNOWLEDGMENTS

This research was sponsored by the Applied Mathematical Sciences Research Program, Office of Energy Research, U.S. Department of Energy under contract DE-AC05-84OR21400 with the Martin Marietta Energy Systems, Inc., and the IBM Bergen Scientific Centre, Bergen Norway.

REFERENCES

1. Landau, L.D. and Lifshitz, E.M., *Electrodynamics of Continuous Media* Pergamon Press, Oxford, 1981.

2. Jaswon, M.A. and Symm G.T. *Integral Equation Methods in Potential Theory and Elastostatics* Academic Press, London, 1977.

3. Mitra, A.K. and Ingber, M.S. Resolving difficulties in the BIEM caused by geometric corners and discontinuous boundary conditions, in *Boundary Elements IX*, (Ed. Brebbia, C.A., Wendland, W.L. and Kuhn G.), pp. 519-532, Proceedings of the 9th Int. Boundary Element Conf., Stuttgart, W. Germany, 1987. Computational Mechanics Pubs., Southampton, 1987.

4. Ingham, D.B. and Kelmanson, M.A. *Boundary Integral Equation Analyses of Singular, Potential, and Biharmonic Problems*, Springer-Verlag, Berlin and New York, 1984.

5. Brebbia, C.A., Telles, J.C.F. and Wrobel, L.C. *Boundary Element Techniques,* Springer-Verlag, Berlin and New York, 1984.

6. Gray, L.J., Giles, G.E. and Bullock, J.S. Progress on boundary element techniques for electroplating simulation, in *Boundary Element Techniques: Applications in Fluid Flow and Computational Aspects* (Ed. Brebbia, C.A. and Venturini, W.S.), pp. 161-173, Computational Mechanics, Southampton, 1987.

Boundary Elements with Macro Shape Functions

Gautam Dasgupta
Department of Civil Engineering and Engineering Mechanics, Columbia University, New York, NY 10027, USA

ABSTRACT

The principles of kinematic continuity constitute the foundation of all spatial discretization techniques. The boundary element method implicitly satisfies all continuity requirements to the appropriate degree. The reason is that the space of the boundary element test functions are tangible continuum solutions. Analogous finite element type shape functions can be viewed as the integrals of the Green functions conjugated with the stipulated constant distribution of the element boundary loads. This notion led to *rational approximations* for the global shape functions. An identical polynomial appears in the denominators of such Padé shape functions pertaining to all nodes irrespective of the shape and size of the object domain. These shape functions provide excellent interpolants to almost all field problems for which a precise field equation of kinematics involve variables of evolutionary type. The morphometric analysis of form and growth of biological objects furnishes a good example. The idealization with the popular isoparametric finite element results as a degenerate case only when the morphology of the interior is of no consequence. The original Courant triangulation procedure is herein extended to arbitrary convex regions to construct the global shape functions.

INTRODUCTION

The boundary element method has been viewed by a number of researchers to be a finite element method with a single consistent element with global shape functions. Refer to Washizu [1, pp. 518 - 537, 617 - 625]. In the conventional form the boundary element basis functions are the Green function solutions. For a system of canonical elliptic partial differential equations, these *flexibility* forms of solution can yield the appropriate eigenfunctions to be employed in the *stiffness* type analysis. For all dimensions and for all degree of elliptic partial differential equations the analytical continuation properties are inherent in these Green functions, Stakgold [2]. Hence the corresponding displacement interpolations necessarily should abide by the maximum and the minimum principles. From the finite element viewpoint this requirement rejects a wider class of the incompatible or non-conforming shape functions.

One-dimensional regions have been modeled with piecewise linear functions which satisfy the second order ordinary differential equation. In two-dimensions a triangular region is the simplest enclosed area with straight line segments. This is then the analog of the line segments of one-dimension. A linear function in two coordinates is still a valid shape function. This idea was proposed by Courant in 1941, [3]. Commercial applications originated with Turner's triangulation to solve problems of structural dynamics which were labeled by Courant as instances of equilibrium and vibrations. Variational methods provided the alternative to the field equations of equilibrium. This constitutes an example of a Ritz projection operator [4] whose solution can be refined to an arbitrary degree according to the Kantorovic weighted residual procedure [5]. However, the intrinsic problems with finite element type local discretization with zero energy mode and boundary incompatibility can be traced to the selection of shape functions which are not the trial functions for the elliptic partial differential equation as demanded by the boundary element procedure[6].

In order to stress the issue of kinematic consistency a problem is selected where the flux balance principle is asserted only in a non-deterministic sense. Thus in the mean stochastic perspective the kinematics would suffice. Such are the problems of analysis of shape and size as encountered in robotic vision, pattern recognition [7] and identification in anatomical growth process for biological entities. Numerical attention will be confined in this paper to analyze a particular dataset which has undergone a considerable amount of biological scrutiny.

The cranio-facial growth analysis of a population of rats is undertaken. Refer to Fig.1. The skull boundary is defined in terms of spatial nodes whose locations were measured by Vilmann, vide acknowledgements. These nodes bear biological significance and constitute a homologous set. In order to define the growth at different time stages a time dependent map can be easily obtained for the boundary nodal points. In conventional finite element isoparametric formulation, shape functions, restricted by the order of polynomials, need to be enforced in order to estimate the distribution of the interior growth. The objection from the morphologists is that the polynomial continuity could indicate negative growth at certain interior locations which are known to display enlargement in size as dictated by the underlying biological process. The problem is more severe in the statistical scenario. Therein a substantial amount of finite element data has to be discarded as outliers. Macro shape functions in the light of the boundary element method are hereby proposed to remedy the situation.

ANATOMICAL TOPOGRAPHY

The notion of *anatomical landmarks* is well accepted in the vocabulary of the biological description for growth and form. The point-wise prescription of geometry is indeed natural to the conventional finite element method. In essence *the anatomical landmarks* form the principal nodes on the boundary for the development of the finite element mesh.

Recent advancement with *higher order* (boundary element type) shape functions permits one to accept **curved element boundaries** with the description in the form of **continuous functions**. This allows the morphologist to prescribe the shape of a biological object with *anatomical elements* without being restricted by discrete landmarks. The boundary shape is very important to depict the biological behavior of the object under study. Such curved boundary elements will not be discussed in this paper. Instead attention will be focused on convex polygons. It should be noted that it is possible to exceed the point-wise (rather limited) description of anatomical landmarks which have a strong flavor of being geometrically isolated singular points. **One can now specify anatomical curves or surface segments.** The algebraic functions, which describe the curved geometry, are **directly** used to generate the Padé finite element interpolations. This is indeed an enhancement in morphometric descriptions of considerable biological significance.

Mathematically speaking, the polygonal boundary description is one *level higher* in sophistication than the conventional isoparametric finite element models. It is thus anticipated that the shortcomings of the triangular or isoparametric quadrilateral finite elements with anatomical landmarks could be circumvented with the proposed model which is constituted by the proposed *rational elements*. The name indicates a Padé rational approximation for the growth-field where the polynomial expressions appear both in the numerator **as well as** in the denominator. In a sense the Padé form is *equivalent* to having *infinite terms* in a Taylor polynomial expansion. This construction makes the polynomial algebra quasi-complete.

The rewarding numerical accuracy with the Padé shape functions only comes with an additional computational complexity. Computer algebra as opposed to **FORTRAN**-based computer arithmetic becomes essential. Programs, like **MATHEMATICA**, refer to [8], become essential in a digital computer environment.

As a side-effect of the rational element construction, an associated *shape space* evolves where the statistical nature of variability in the form and growth can be viewed unambiguously in a frame-invariant fashion.

PADÉ ELEMENTS

Computer Algebra programs are particularly suited for manipulating polynomials, rationalizing expressions and evaluating derivatives of Taylor and Padé sequences. Thus, the algebraic chores pertaining to the finite element formulation of problems related to equilibium and vibrations, i.e. the construction of local stiffness and mass matrices, do not pose any difficulty when the boundary geometry exceeds the traditional complexity bounds of isoparametric (quadrilateral) elements.

In this paper attention is focused on furnishing closed form expressions for shape functions associated with polygons with arbitrary number of sides. The same procedure holds for polyhedra and convex shaped domains in higher dimensional (hyper) spaces. It is anticipated that with the emergence of the academic workstations, like the *NEXT Computer System*, where the *Symbolic Manipulator* **MATHEMATICA** [8] is furnished within the standard operating system, a number of practical discretization schemes of better mathematical precisions could evolve. Literally a hundred pages of algebraic substitution does no longer stand as the *de facto* definition of intricacy.

GENERAL FORMULATION

Consider the following **elliptic** partial differential equation :

$$\mathcal{L}[\{u_1,\ldots,u_m\},\{x_1,\ldots,x_n\}] = f[\{x_1,\ldots,x_n\}]$$

valid in a domain $\mathcal{D}$ with boundary $\mathcal{B}$. In the above equation, $\mathcal{L}$ is a linear operator governing m variables $u's$ subjected to a forcing function f with $x's$ constituting the independent coordinates.

The *finite element method* is very apt in dealing with arbitrary shapes of the domain $\mathcal{D}$ with piecewise approximation of the boundary $\mathcal{B}$. *In general*, attention is paid on the nature of the solution on a simplified region with p nodes in order to construct a set of basis functions $N_1,\ldots,N_p$. Traditionally a polynomial expression is chosen with p unknowns.

On the other hand, the *Boundary Element Method* more accurately selects the $u's$ as a linear combination of *Green's Functions* **G** defined by:

$$\mathcal{L}\mathbf{G} = \delta[\{x_1,\ldots,x_n\} - \{x_1^0,\ldots,x_n^0\}]$$

Experience shows that usually the Green's function is best expressed in terms of higher order transcendental functions. Obviously a Padé type *rational approximation* will suit better than a polynomial Taylor form. This observation prompted the present formulation.

Derivation of a First Order Element

A convex polygon of p sides is considered. It is possible then to define uniquely the equation of the ith side in the form:

$$g_i(x,y) = 1 + a_i x + b_i y = 0,$$

with respect to the $x - y$ frame. On the plane for any interior point (x^0, y^0) the following inequality would hold:

$$g_i(x^0, y^0) > 0$$

Let the nodes and the sides be so numbered that the sides with number i and $i+1$ intersect at node i.

The *first order shape function*, N_i, associated with the node i, is hereby defined to be an interpolation of *the simplest form* which vanishes *all along* the boundary except on line segments i and $i+1$ and has a unit value at the node i. It is then possible to obtain an expression for $N_i(x,y)$ in the Lagrange interpolation form:

$$N_i(x,y) = \kappa \prod_{j \neq i, i+1} g_j(x,y) \tag{1}$$

and κ is such that

$$\sum_{j=1}^{p} N_i(x,y) = 1$$

It should be noted that the aforementioned construction of the global shape function is merely the generalization of the Courant triangulation procedure as extended to convex polygons. The advantage of the proposed technique is that one does not explicitly evaluate the coefficients of a polynomial in terms of the nodal coordinates to obtain N_i. This is similar to the Lagrange interpolation idea. The problem with the incompleteness feature of polynomials in Banach spaces is also circumvented.

As an illustration of Eq. (1), symbolically the first shape function of a pentagon can be cited to be:

$$N_1 = \frac{g_3 g_4 g_5}{(g_3 g_4 g_5 + g_4 g_5 g_1 + g_5 g_1 g_2 + g_1 g_2 g_3 + g_2 g_3 g_4)} \tag{2}$$

Equation (2) in the familiar **rational** form, where polynomials appear as numerator and denominator, is a Padé expression. The computed shape functions for equilateral hexagonal and octagonal elements plotted as displayed in Figs. 2(a) and 2(b).

It is obvious that for an interior point (x^0, y^0) of the **convex** domain

$$1 > N_i(x^0, y^0) > 0$$

This maximum/minimum principle is an absolutely necessary feature for a shape function pertaining to an elliptic partial differential equation.

Evaluation of $N_i(x,y)$ and their derivatives even for the Green strain expressions with large rotation is virtually a *trivial* problem when one uses a suitable computer algebra program such as the **Mathematica**.

DERIVATION OF HIGHER ORDER MACRO SHAPES

The simplest form of the shape function which yielded the *first order element* shows that, for the shape function $N_i(x,y)$, the distribution along the line $g_i(x,y) = 0$ follows the ratio:

$$\frac{g_{i-1}(x,y)}{g_{i-1}(x,y) + g_{i+1}(x,y)} \tag{3}$$

with the stipulation that g_0 is g_p. The higher order element will approximate $\{N_i(x,y)\}$ when the distributions of the fundamental (*Boundary*

Element type) responses are to be accommodated. It is a routine activity for the computer algebra programs to effect Padé expansions in terms of the bases functions $\{g_i(x,y)\}$. For example, if the distribution along the boundary segments $\{i\}$ is required to follow a prescribed profile $\{l_i(x,y)\}$ then a set of functions $\{f_i(x,y)\}$ can be evaluated uniquely, based on a specified numerical error bound, such that the Green's form of distributions are met. Then one simply replaces the set $\{g_i(x,y)\}$ by $\{f_i[g_i(x,y)]\}$ in the expressions of the *first order shape functions* $N_i(x,y)$ to obtain the proposed higher order ones. It should be mentioned that during the execution of the Padé approximation the constraint $\mathbf{f_i(0) = 0}$ has to be satisfied. An example of $f_i(z)$ to be Log(z+1) is illustrated in reference [9].

DERIVATION OF A GENERALIZED CONVEX ELEMENT

We have an intuitive notion of convexity. Topologically, a domain is convex if for two interior points the connecting straight line segment lies entirely within the domain. In this presentation we shall invoke more of an algebraic topological argument rather than a Euclidian geometric one. A *generalized convex domain*, in n-dimensions, is defined with respect to a set of *equations* $\{g_i(x_1,\ldots,x_n)\}$ if for each internal point $(x_1^0,\ldots,x_n^0)$ the positiveness of $\{g_i(x_1^0,\ldots,x_n^0)\}$ is guaranteed. Refer to Fig. 2(c). In there the region **D** is not convex in the Euclidean sense. However, according to the following equations:

$$g_1 : 1+x-y; \quad g_2 : 1+x+y; \quad g_3 : y^2-4x$$

the domain **D** qualifies to be a *generalized convex domain*. It is evident that the aforementioned expressions for the shape functions are indeed valid.

DERIVATION OF A COMPACT L'HOPITAL FORM

One of the outstanding features of the computer algebra programs is their pattern recognition capability. This provides a very flexible tool to define mathematical entities beyond the strict substitute type enumeration. For example a limiting value of $\frac{Sin[x]}{x}$ could be mathematically sought as x becomes indefinitely close to 0. In there the L'Hopital's rule will be applied **automatically** without any special request from the user. This situation indeed exceeds the expectation of a traditional **FORTRAN** user. In designing the computational strategy for elements with a large number of boundary segments p a simple alteration is proposed to enhance the symbolic computations by several orders of magnitude. One can rewrite Eq.(1) in the following form:

$$\frac{1}{N_i(x,y)} = \kappa \sum_{j=i,i+1} \frac{1}{g_j(x,y)} \qquad (4)$$

For a polygon with p sides, for example, instead of a product $(p-2)$ terms in the numerator, we will have only two algebraic expressions to deal with. This saves computational time. Furthermore, an element with p sides would require computation time proportional to p as in equation (4) not to p^2 of Eq. (1).

NUMERICAL EXAMPLES

The growth data is shown in Fig. 1. The Green strain tensor is plotted as contours to depict the growth distribution in Fig. 3. The maximum and minimum (strains) growth rates are shown in Figs. 3(a) and 3(b), respectively. The angle between the maximum growth rate vector and the principal axis marked $x-$ in Fig. 1(b) is described in Fig. 3(c). The continuity in the computed growth rates over the entire macro element of Fig. 1 (b), conforms to the biologically significant processes which underlie the growth phenomenon.

ACKNOWLEDGMENTS

This research was sponsored by:

1) Alexander von Humboldt Foundation, West Germany
2) National Institute of Health, U.S.A.
3) National Center for Earthquake Engineering Research, U.S.A.

The growth data was measured by Prof. Dr. Henning Vilmann, Royal College of Dentistry, Copenhagen, Denmark.

REFERENCES

[1] Washizu, K., Variational Methods in Elasticity and Plasticity, third edition, Pergamon Press, NY, 1982.

[2] Stakgold, I., Green's Functions and Boundary Value Problems, John Wiley, New York, 1979.

[3] Courant, R., "Variational Methods for the Solution of Problems of Equilibrium and Vibration " , Bull. Am. Math. Soc., 49, 1-23, 1943.

[4] Kurpel, N. S. Projection-Iterative Methods for Solutions of Operator Equations, AMS publication, Providence, R.I., 1976.

[5] Kantorovic, L. V. Functional Analysis and Applied Mathematics National Bureau of Standards, Washington, D. C., Report No. 1509, 1952.

[6] Zienkiewicz, O. C. and Taylor, R. L. The Finite Element Method, Fourth Edition, Volume 1, McGraw-Hill, 1989.

[7] Proceedings of IEEE International Conference on Neural Networks, San Diego, California, 1988.

[8] Wolfram, S. Mathematica A System for Doing Mathematics by Computer, Addison Wesley, NY, 1988.

[9] Publico, A., and Dasgupta, G., "Hexagonal and Octagonal Finite Elements: Closed Form Expressions for Shape Functions",*Proceedings of CANCAM 89*, Department of Civil Engineering, Carleton University, Ottawa, Ontario, Canada K1S 5B6.

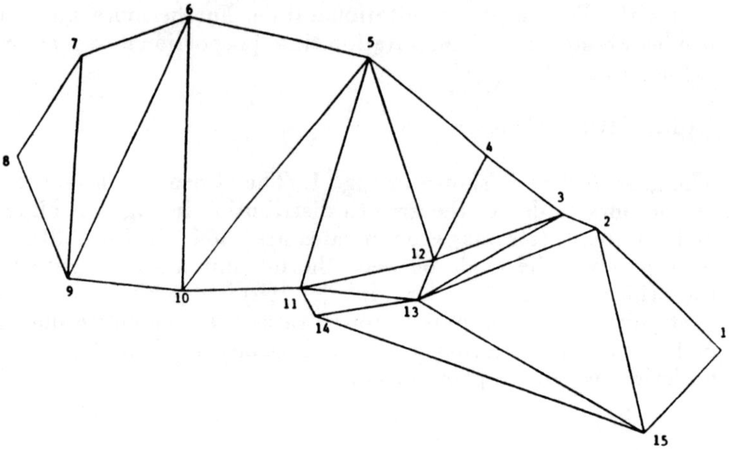

1.(a) Finite Element Triangulation

NODE	5	6	7	8	9	10	11	12
X	28.34	-16.36	-43.49	-48.72	-33.17	-1.25	29.79	58.28
Y	27.90	29.09	19.65	-6.56	-28.73	-27.86	-23.22	-8.97

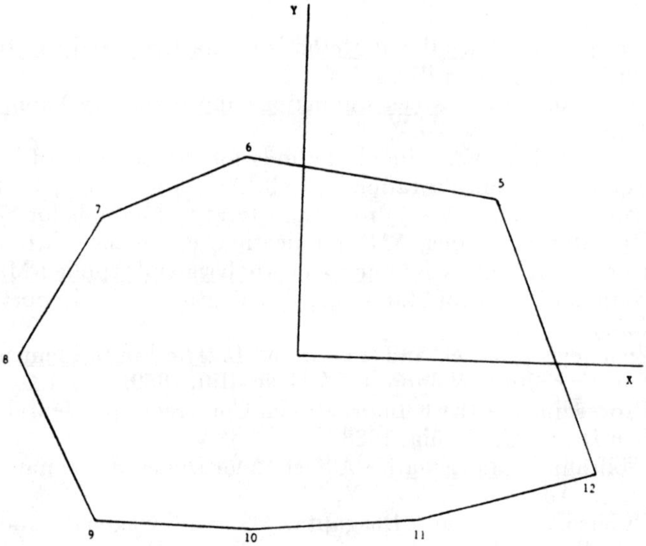

1.(b) Macro Element with Characteristic Coordinates

Fig. 1: Rat Skull

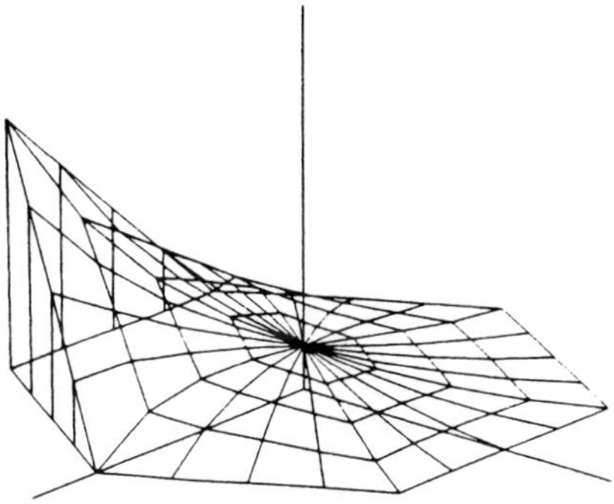

2.(a) Shape Function for Hexagonal Macro Element

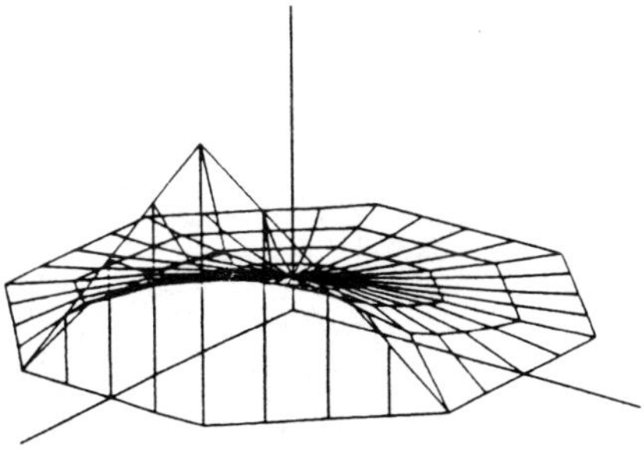

2.(b) Shape Function for Octagonal Macro Element

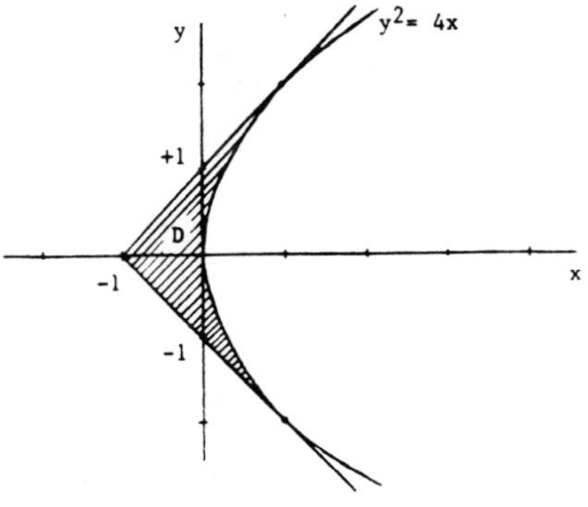

2.(c) Generalized Convex Region

Fig. 2: Macro Elements

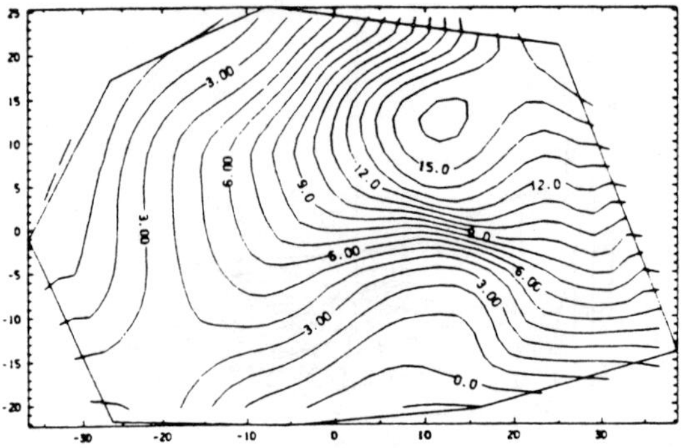

3.(a) Green's First Principal (strain) Growth Rate

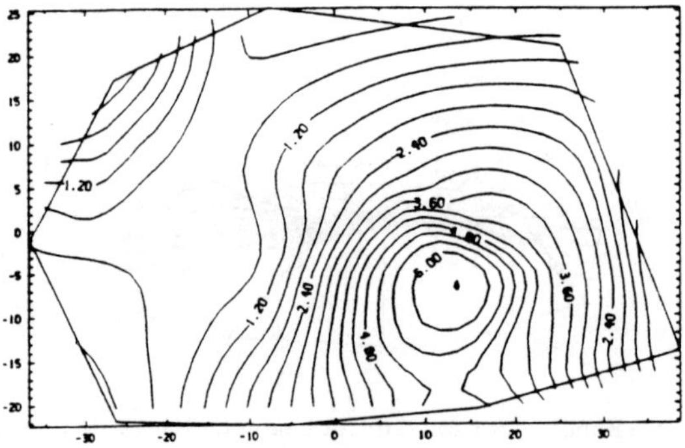

3.(b) Green's Second Principal (strain) Growth Rate

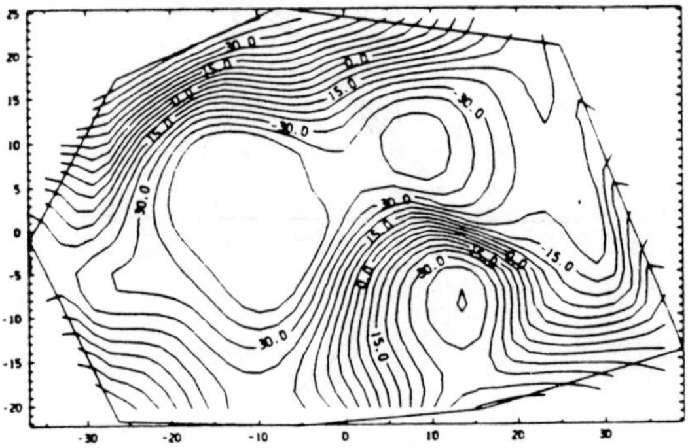

3.(c) Orientation of Green's Principal (strain) Growth Rate (in degrees)

Fig. 3: Spatial Growth Rate in a Rat Skull

Iterative Solvers for BEM Systems of Equations

F.C. Araújo and W.J. Mansur
Civil Engineering Department, COPPE/UFRJ, C.P. 68506, 21945 Rio de Janeiro, Brazil

ABSTRACT

This paper discussion is concerned with the application of iterative solvers to find solutions of algebraic linear systems of equations originated from the Boundary Element Method.

A brief review related to previous approaches reported in the literature is first presented. Subsequently, the authors' numerical experiments are described.

Two numerical examples are analysed, in order to compare the efficiency of the iterative solvers studied with that of traditional Gauss Elimination procedure.

INTRODUCTION

Since the Boundary Element Method started being used as an alternative numerical technique, direct Gauss elimination procedure has been used to solve the system of equations generated. This solution technique, when associated with pivoting, is quite trustful having been used in a large number of applications, with no failure reported so far in the literature. For this reason it has been preferred by most researchers and engineers, and is the technique used in most (if not all) of the commercial BE codes available. However Gauss elimination can become very expensive when employed to solve problems with large number of unknowns as its solution time for nonsymmetric systems of equations is proportional to N^3, being "N" the order of the system. In order to reduce Boundary Element analysis costs, which for large problems is directly related to the system of equation solution computer time, researchers started studying iterative solvers as a possible cheaper alternative solution method, as can be seen in the works developed by Doblare [1], Parreira [2], Bettess [3], Hencis and Muller [4]. Although none of these works have achieved a cheap and trustful solution algorithm, they are very important as they called attention of the Boundary Element Scientific Community that we must not accept the high costs of Gauss elimination technique passively.

In this paper part of a research developed by the authors at COPPE-UFRJ on iterative solvers for BEM system of equations is reported. A more comprehensive study can be found in Araujo [5].

The iterative methods described here are used to find solutions of

$$\underset{\sim}{A}\ \underset{\sim}{U} = \underset{\sim}{b} \tag{1}$$

by employing the following recursive procedure

$$\underset{\sim}{U}^{n+1} = \underset{\sim}{U}^{n} + \lambda_n \underset{\sim}{P}^{n} \tag{2}$$

ITERATIVE SOLVERS

There exist a large number of iterative techniques which can be employed to solve nonsymmetric systems of equations, for a comprehensive study the reader is referred to Jennings [6], Hageman and Young[7], Axelsson [8], Saad [9].

In this work, the solvers selected were those which did not impose any "a priori" restriction to the system matrix, that is solvers for positive definite matrices, positive real matrices, etc were not considered. Studies taking into consideration special features of BEM system matrices have been postponed until more is known about their particular characteristics.

The first algorithm studied (CG-TWO) is the standard conjugate gradient acceleration adapted to nonsymmetric matrices [6,7].

The second algorithm studied is the biconjugate gradient acceleration (bi-CG) as described by Saad [9], which can be seen as an acceleration procedure over the RF method described in reference [7]. In the present work the acceleration procedure presented in reference [9] is generalized to be used to accelerate other basic iterative methods. Particularly, this acceleration procedure has been applied over Jacobi basic iterative method described in reference [7]. The iterative scheme which resulted from the procedure described is referred to as J/bi-CG.

CG-two and J/bi-CG algorithms are as follows:

CG-TWO (Two-Term Conjugate Gradient Method)

$\underset{\sim}{U} = \underset{\sim}{U}^0$

$\underset{\sim}{r} = \underset{\sim}{b} - \underset{\sim}{A}\underset{\sim}{U}$

$\underset{\sim}{\delta} = \underset{\sim}{A}^T \underset{\sim}{r}$

$DELT\phi = \underset{\sim}{\delta}^T \underset{\sim}{\delta}$

IF $DELT\phi \leqslant TOL \Rightarrow$ STOP

$\underset{\sim}{P} = \underset{\sim}{\delta}$

R: $\underset{\sim}{h} = \underset{\sim}{A}\ \underset{\sim}{P}$

$\underset{\sim}{h}_1 = \underset{\sim}{A}^T \underset{\sim}{h}$

$\lambda = DELT\phi / \underset{\sim}{P}^T \underset{\sim}{h}_1$

$\underset{\sim}{U} = \underset{\sim}{U} + \lambda \underset{\sim}{P}$

$\underset{\sim}{\delta} = \underset{\sim}{\delta} - \lambda \underset{\sim}{h}_1$; $DELT1 = \underset{\sim}{\delta}^T \underset{\sim}{\delta}$

IF $DELT1 \leqslant TOL \Rightarrow$ STOP

$\alpha = DELT1/DELT\phi$; $DELT\phi = DELT1$

$\underset{\sim}{P} = \underset{\sim}{\delta} + \alpha \underset{\sim}{P}$

GO TO R

J/bi-CG (Jacobi - Biconjugate Gradient Acceleration)

In the algorithm described below Q is the Jacobi basic iterative method splitting matrix, that is, all of the coefficients of Q are null except those on the diagonal, which are equal to the diagonal terms of the system matrix A shown in equation (1).

$$\underset{\sim}{U} = \underset{\sim}{U}^0$$

$$\underset{\sim}{r} = \underset{\sim}{b} - \underset{\sim}{A}\,\underset{\sim}{U}$$

$$\underset{\sim}{\delta} = \underset{\sim}{Q}^{-1}\,\underset{\sim}{r}$$

$$\mathrm{DELT\phi} = \underset{\sim}{\delta}^T\,\underset{\sim}{\delta}$$

$$\text{IF DELT}\phi \leqslant \text{TOL} \Rightarrow \text{STOP}$$

$$\underset{\sim}{P} = \underset{\sim}{\overline{P}} = \underset{\sim}{\overline{\delta}} = \underset{\sim}{\delta}$$

R: $$\underset{\sim}{h} = \underset{\sim}{A}\,\underset{\sim}{P}$$

$$\underset{\sim}{h_1} = \underset{\sim}{Q}^{-1}\,\underset{\sim}{h}$$

$$\lambda = \mathrm{DELT\phi}/\underset{\sim}{P}^T\,\underset{\sim}{h}_1$$

$$\underset{\sim}{U} = \underset{\sim}{U} + \lambda\,\underset{\sim}{P}$$

$$\underset{\sim}{\delta} = \underset{\sim}{\delta} - \lambda\,h_1 \;;\; \mathrm{DELT1} = \underset{\sim}{\delta}^T\,\underset{\sim}{\delta}$$

$$\text{IF DELT1} \leqslant \text{TOL} \Rightarrow \text{STOP}$$

$$\underset{\sim}{h} = \underset{\sim}{Q}^{-T}\,\underset{\sim}{P}$$

$$\underset{\sim}{h_1} = \underset{\sim}{A}^T\,\underset{\sim}{h}$$

$$\underset{\sim}{\overline{\delta}} = \underset{\sim}{\overline{\delta}} - \lambda\,h_1$$

$$\mathrm{DELT1} = \underset{\sim}{\overline{\delta}}^T\,\underset{\sim}{\delta}$$

$$\alpha = \mathrm{DELT1}/\mathrm{DELT\phi} \quad ; \quad \mathrm{DELT\phi} = \mathrm{DELT1}$$

$$\underset{\sim}{P} = \underset{\sim}{\delta} + \alpha\underset{\sim}{P}$$

$$\underset{\sim}{\overline{P}} = \underset{\sim}{\overline{\delta}} + \alpha\underset{\sim}{\overline{P}}$$

GO TO R

APPLICATIONS

The system of equations obtained from a BEM analysis reads:

$$\underset{\sim}{H}\,\underset{\sim}{U} = \underset{\sim}{G}\,\underset{\sim}{P} \tag{3}$$

which according to the boundary conditions can be reordered and written as:

$$\underset{\sim}{A}\,\underset{\sim}{X} = \underset{\sim}{Y} \tag{4}$$

The matrix of the final system of equations (matrix $\tilde{A}$) is formed by coefficients of matrices $\tilde{H}$ and $\tilde{G}$ of equation (3) which may have different magnitude. This is not much of a problem in what concerns Gauss elimination costs, however it can influence the solution obtained via iteractive solvers; therefore it is desirable to perform coefficient scaling. Guidelines for potential problems are given by Bettess [3].

The applications carried out in this work concern two dimensional linear elasticity problems. In order to make Gauss elimination CPU time shorter, pivoting was not employed. The iterative procedures used, solved linear systems of equations generated according to the procedures described in reference [10]. A simplified coefficient scaling has been used, which is the same employed in reference [11], where equation (3) is used as follows:

$$\tilde{H}\,\tilde{U} = \tilde{G}'\,\tilde{P}' \tag{5}$$

where,

$$\tilde{G}' = C_8\,\tilde{G}$$

$$\tilde{P}' = (1/C_8)\tilde{P}$$

$$C_8 = 2G/(1-\nu)$$

being G the shear modulus and ν the Poisson coefficient.

Coefficient scaling is a subject which must be further studied.

It is also important to notice that non-nodal-collocation elements [12] were used in order to deal with traction discontinuities. In this type of element the collocation point is dislocated towards its interior by a distance which in reference [10] was taken to be 25 per cent of the element length. At the beginning of the study the use of iteractive solvers together with non-nodal-collocation elements was a subject of concern. However no disturbing effect has been noticed.

Linear elements were used on the BEM analysis described next.

<u>Square Plate</u>

This plane stress example consists of a square plate under biaxial load. Initial discretization, boundary conditions and material properties are shown in figure 1. Non-nodal-collocation elements were used adjacent to the corner nodes.

In order to check the performance of CG-two and J/bi-CG against Gauss elimination, the mesh shown in figure 1 was successively subdivided. Results of the various analyses are shown in figure 2 and table 1.

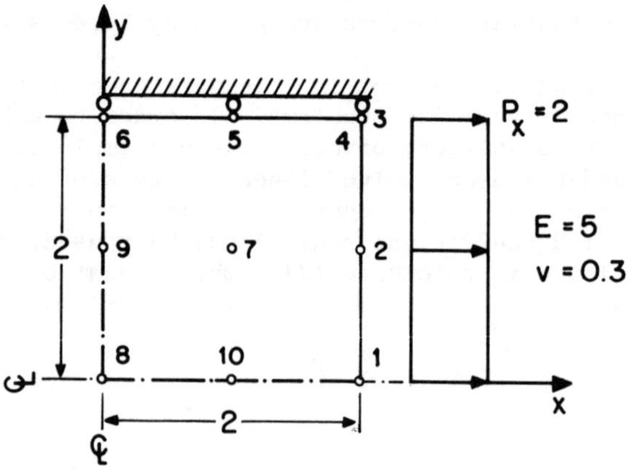

Figure 1. Square Plate

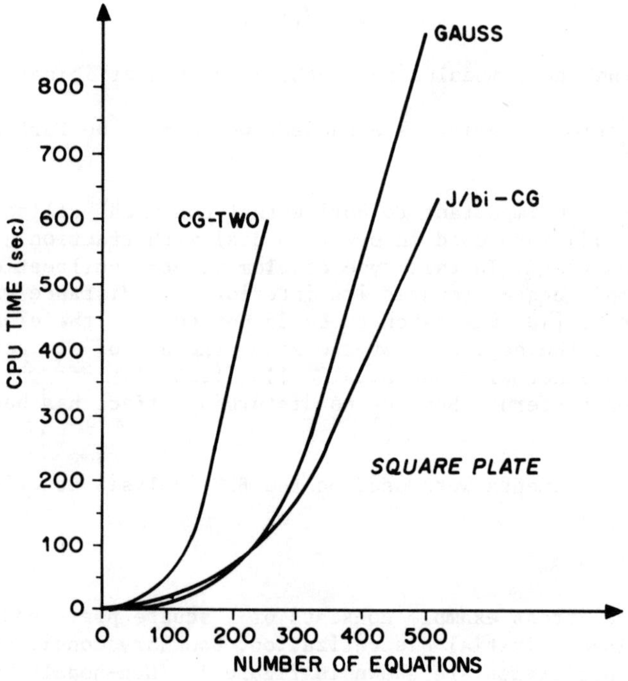

Figure 2. CPU time versus number of equations for square plate

	SOLVER / NUMBER OF EQUAT.	GAUSS	CG-TWO	J/bi-CG
NUMBER OF ITERATIONS	68	-	96	32
	100	-	148	35
	204	-	283	44
	300	-	-	51
	500	-	-	66
CPU TIME (sec)	68	2.	16.	5.
	100	7.	52.	12.
	204	61.	417.	65.
	300	194.	-	162.
	500	884.	-	572.

Table 1. Number of iterations and CPU time for square plate

Hole in a Plate

In this example the plane stress problem of an infinite plate with a hole is studied. Discretization, boundary conditions and physical properties are depicted in figure 3. Results for this analysis are shown in figure 4 and table 2.

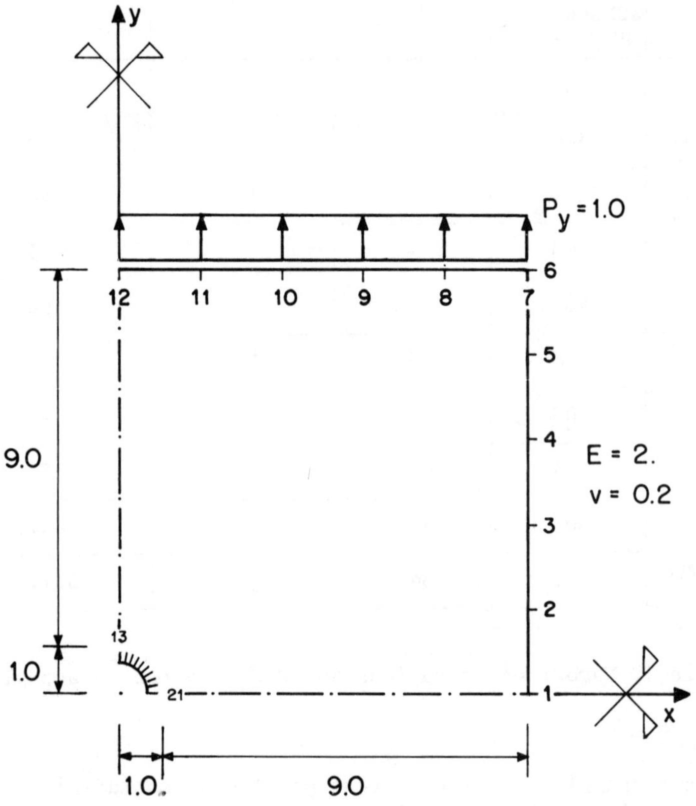

Figure 3. Plate with a hole

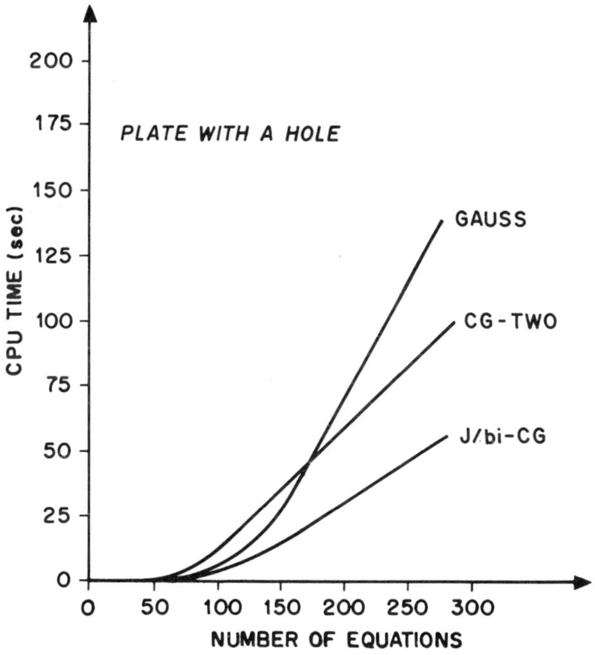

Figure 4. CPU time versus number of equations for plate with a hole

	SOLVER / NUMBER OF EQUAT.	GAUSS	CG-TWO	J/bi-CG
NUMBER OF ITERATIONS	42	-	25	13
	98	-	31	14
	134	-	47	20
	258	-	35	22
CPU TIME (sec)	42	0.5	1.5	0.8
	98	6.	11.	5.
	134	17.	31.	13.
	258	122.	86.	51.

Table 2. Number of iterations and CPU time for hole in plate

CONCLUSIONS

Results obtained in this paper and summarized in tables 1 and 2 and figures 2 and 4 show that iterative solvers can be used for finding solutions of BEM systems of equations.

The procedure denominated CG-TWO proved to be less efficient than Gauss elimination, this conclusion being just a confirmation of previous researchers' studies.

The procedure denominated J/bi-CG had a much better performance than CG-TWO, and was more efficient than Gauss elimination for medium sized systems of equations (250 equations). As the order of the system increased, J/bi-CG became more economical, quite large savings being expected if this solver is used in industrial applications.

Finally, it is important to mention the importance of working out an effective coefficient scaling technique in order to make iterative solvers for BEM analysis even more efficient. This becomes clear if it is observed that figures 2 and 4 show that in the first and second application J/bi-CG became cheaper than Gauss elimination for 250 and 98 unknowns respectively. The reason for this is that in the first example the system matrix is formed by coefficients of H and G whereas in the second one, it is formed by $\underset{\sim}{H}$ matrix coefficients only.

REFERENCES

1. Doblaré, M. (1981). Formulación Tridimensional del Método de los Elementos de Contorno con Interpolación Parabólica (in spanish), Ph.D. Thesis Universidad Politécnica, Madrid.

2. Parreira, P. (1987). Análise do Erro no Método dos Elementos de Fronteira em Elasticidade (in Portuguese), Ph.D. Thesis, Universidade Técnica de Lisboa, Portugal.

3. Bettess, J.A. (1984). Solution Techniques for Boundary Integral Matrices, Proceedings of the International Conference on Numerical Methods for Transient and Coupled Problems, Venice, July.

4. Mullen, R.L. and Rencis, J.J. (1987). Iterative Methods for Solving Boundary Element Equations, Computers & Structures, Vol. 25, No. 5, pp. 713-723.

5. Araújo, F.C. (1989). Técnicas Iterativas para a Solução de Sistemas de Equações Lineares Oriundos do Método dos Elementos de Contorno (in Portuguese), Ms.C. Thesis COPPE-UFRJ, Rio de Janeiro.

6. Jennings, A. (1977). Matrix Computation for Engineers and Scientists, John Wiley, New York.

7. Hageman, L.A. and Young, D.M. (1981). Applied Iterative Methods, Academic Press, New York.

8. Axelsson O. and Barker V.A. (1984). Finite Element Solution of Boundary Value Problems, New York.

9. Saad, Y. (1982). The Lanczos Biorthogonalization Algorithm and Other Oblique Projection Methods for Solving Large Unsymmetric Systems, Siam J. Numer. Anal., Vol. 19, No. 3, pp. 486-506.

10. Neves, A.C. (1988). Resolução de Problemas Viscoelásticos Utilizando o Método dos Elementos de Contorno (in Portuguese), Ms.C. Thesis COPPE-UFRJ, Rio de Janeiro.

11. Brebbia, C.A., Telles, J.C.F. and Wrobel, L.C. (1984). Boundary Element Techniques: Theory and Applications in Engineering, Springer Verlag, Berlin and New York.

12. Marques, E. and Mansur, W.J. (1987). Coupling of Boundary and Finite Element Methods: Application to Potential Problems, Proceedings of VIII MECOM Conference, Rio de Janeiro.

Semianalytic Boundary Element Solution of Time Dependent Partial Differential Equations

M.M. Aral and Y. Tang

School of Civil Engineering, Georgia Institute of Technology, Atlanta, Georgia 30332, USA

ABSTRACT

A numerical solution is developed for the analysis of parabolic partial differential equations using semianalytic time integration procedures and the boundary element method (BEM). The secondary reduction method (SR-BEM) developed earlier, is an integral part of this computation process. Compared to finite difference or other time integration procedures, the semianalytic computation process yields more efficient and accurate results. The proposed method is tested here for classic two dimensional heat transfer problems. Numerical results obtained are verified using analytical procedures and presented comparatively with other boundary element solutions.

INTRODUCTION

The objective of this paper is to illustrate the use of semianalytic time integration procedures in boundary element solution of time dependent heat transfer problems. In the literature, several other procedures which utilize BEM were suggested to solve parabolic partial differential equations. These may include finite difference approximations of the time derivative, Curran et al. [5], Laplace transforms, Rizzo and Shippy [6], and time dependent fundamental solutions, Chang et al. [4]. A detailed discussion of these techniques can be found in standard text books on boundary element methods.

Recently, a secondary reduction boundary element procedure was suggested by Aral and Tang [1,2,3], which yields efficient algorithms for the solution of time dependent parabolic partial differential equations. In earlier applications of SR-BEM, finite difference time integration procedures were used to integrate the solution in temporal dimension. The purpose here is to demonstrate the use of SR-BEM formulation in reducing the final boundary element matrix equations to a form suitable for the application of semianalytic time integration procedures. As demonstrated in this paper, this process may also be employed in reducing the final matrix equations developed by other procedures such as Dual Reciprocity Boundary Element process (DR-BEM), Wrobel, L. C. et. al. [8]. The main motivation in use of semianalytic procedures lies in the fact that the semianalytic method is applied after the spatial solution has been performed using boundary element procedures. Since analytical procedures are used in the time integration algorithms, the spatial errors should not propagate in the temporal dimension, Umari and Gorelick [7]. We also know that, for most problems governed by parabolic partial differential equations, fundamental solutions can be described for exact spatial representation using boundary element approach which minimizes spatial errors in numerical solution. Thus, the use of semianalytic procedures along with the boundary element method promise the best combination in solution of time dependent problems. The formulation and the advantages and disadvantages of this computation is discussed in this paper.

BOUNDARY ELEMENT FORMULATION

The general nondimensional form of the governing equation for a time dependent heat conduction problem in a two-dimensional homogeneous region can be given as,

$$\frac{\partial \phi^2}{\partial x^2} + \frac{\partial \phi^2}{\partial y^2} = \frac{\partial \phi}{\partial t} \tag{1}$$

In Equation (1), ϕ is the temperature and (x,y,t) are the dimensionless space and time coordinates. In order to describe a physical problem properly, appropriate Dirichlet and/or Neuman boundary conditions and initial conditions should also be specified. The general boundary element formulation of Equation (1) can be given as follows. We chose the fundamental solution as,

$$G = \frac{1}{2\pi} \text{ Ln } (\bar{r}_i) \tag{2}$$

where the function $\bar{r}_i = (r - r_i)$ is defined as the Euclidian distance between the source point (i) and any field point. Given Equation (2), the weighed integral of Equation (1) can be given as,

$$\int_\Omega G \vec{\nabla}^2 \phi \, d\Omega = \int_\Omega G \frac{\partial \phi}{\partial t} d\Omega \tag{3}$$

where $\vec{\nabla}$ is the two-dimensional gradient operator and Ω defines the domain of integration. Applying Green's second identity, Equation (3) can be written as,

$$\lambda_i \phi(r_i, t) - \int_{\Gamma} [\phi \frac{\partial G}{\partial n} - G \frac{\partial \phi}{\partial n}] \, d\Gamma = \int_{\Omega} G \frac{\partial \phi}{\partial t} \, d\Omega \qquad (4)$$

where Γ represents the boundary of the domain Ω and λ_i is the shape coefficient. Idealization of the boundary of the domain as linear segments, approximation of the dependent variable and its normal derivative within these boundary elements, and numerical integration of boundary element integrals of Equation (4) lead to the standard boundary element matrix form of the left hand side of Equation (4), which is in boundary only form. The right hand side of Equation (4) can also be reduced to a boundary only form using SR-BEM or DR-BEM approaches as described in references [1,2,3,8] which will not be repeated here. The final matrix equations can be written as,

$$[R] \{\frac{\partial \phi}{\partial n}\} - [S] \{\phi\} = [M] \{\frac{\partial \phi}{\partial t}\} \qquad (5)$$

SEMIANALYTIC TIME INTEGRATION

Equation (5), which may be obtained utilizing SR-BEM or DR-BEM approach, can now be reduced to a form suitable for semianalytic time integration using again SR-BEM concepts. According to SR-BEM procedure the dependent variable and temporal derivatives may be approximated at any point inside and at the boundary of the domain by a set of k global coordinate functions multiplied by unknown parameters which are functions of time.

$$\phi_i = \sum_{j=1}^{k} f(\bar{r}_j) \, \beta_j(t) \qquad (6)$$

$$\frac{\partial \phi_i}{\partial t} = \sum_{j=1}^{k} f(\bar{r}_j) \, \dot{\beta}_j(t) \qquad (7)$$

Using these definitions normal derivatives can also be defined as follows,

$$\frac{\partial \phi_i}{\partial n} = \sum_{j=1}^{k} \frac{\partial f(\bar{r}_j)}{\partial n} \, \beta_j(t) \qquad (8)$$

where $f(\bar{r}_j)$ is a function defined for any field point in terms of Euclidian distance function defined earlier. In Equations (6), (7) and (8), the subscript j is used for k reference source points. The choice of k reference source points is totally arbitrary. Although other possibilities exist, Aral and Tang [1], the following is defined for $f(r_j)$ in this study where r_o is the maximum distance between any two nodes.

$$f(\bar{r}_j) = e^{-(\bar{r}_j/r_o)} \tag{9}$$

The following matrix operations may now be performed on the matrices of Equation (5). From (6), (7) and (8),

$$\{\phi\} = [F]\ \{B\}\ ;\ \{\frac{\partial\phi}{\partial t}\} = [F]\ \{\dot{\beta}\}\ ;\ \{\frac{\partial\phi}{\partial n}\} = [B]\ \{\beta\} \tag{10}$$

where [F] and [B] are (kxk) matrices obtained from the exponential function defined in (9). Given boundary conditions of the problem, Equation (5) can be written as,

$$[A] \begin{Bmatrix} \phi \\ \frac{\partial\phi}{\partial n} \end{Bmatrix} + \{C\} = [M]\ \{\frac{\partial\phi}{\partial t}\} \tag{11}$$

where the matrix [A] contains the proper coefficients of the unknown variables $\{\phi\}$ or $\{\partial\phi/\partial n\}$ for each node, and the vector {C} contains the boundary condition terms. Using the definitions given in Equation (10), Equation (11) can be written as,

$$[A]\ [FB]\ \{\beta\} + \{C\} = [M]\ \{\frac{\partial\phi}{\partial t}\} \tag{12}$$

where the matrix [FB] is defined as,

$$\underset{(k\times k)}{[FB]} = \begin{bmatrix} f_{11} & f_{12} & \cdots\cdots & f_{1k} \\ f_{21} & f_{22} & \cdots\cdots & f_{2k} \\ f_{\ell 1} & f_{\ell 2} & \cdots\cdots & f_{\ell k} \\ b_{(\ell+1)1} & b_{(\ell+1)2} & \cdots\cdots & b_{(\ell+1)k} \\ b_{k1} & b_{k2} & \cdots\cdots & b_{kk} \end{bmatrix} \qquad (13)$$

That is, the matrix [FB] contains the proper coefficients obtained from Equation (10) according to the ordering present in the right hand side vector of Equation (11).

Using the inverse definitions of Equation (10), Equation (12) can now be reduced to a form suitable for analytic computations in the temporal dimension.

$$[A]\ [FB]\ [F]^{-1}\{\phi\} + \{C\} = [M]\ \{\frac{\partial \phi}{\partial t}\} \qquad (14)$$

Let,

$$[D] = [A]\ [FB]\ [F]^{-1} \qquad (15)$$

$$[E] = [M]\ [F]^{-1} \qquad (16)$$

$$[P] = [D]^{-1}[M]\ [F]^{-1} \qquad (17)$$

$$\{\bar{C}\} = [D]^{-1}\{C\} \qquad (18)$$

Than,

$$[P]\ \{\frac{\partial \phi}{\partial t}\} = \{\phi\} + \{\bar{C}\} \qquad (19)$$

For a time invariant vector $\{\bar{C}\}$, Equation (19) may now be solved explicitly using analytic procedures. This solution can be given as,

$$\{\phi(t)\} = e^{-[P]t}(\{\phi_o\} + \{\bar{C}\}) - \{\bar{C}\} \qquad (20)$$

where $\{\phi_o\}$ is a vector of initial conditions. The term $e^{-[P]t}$ is referred to as matrix exponential and is a matrix of same dimensions as [P] which may be obtained from the eigenvalues and eigenvectors of [P]. This computation process, as used in this study, is independent of boundary element procedures described above, and is explained in detail in Umari and Gorelick [7], which will not be repeated here.

NUMERICAL EXAMPLES

The numerical procedure described above is tested on two classical heat transfer problems for which analytical solutions are available. The domain of the solution region is idealized as shown in Figure 1. In the first problem initial temperature distribution within the domain is chosen as zero and at time zero the temperature at the left hand side boundary is increased to one while keeping the right hand side boundary temperature at zero. In the second problem initial temperature distribution in the domain is chosen as one and at time zero both left and right side boundary temperatures are reduced to zero. In both of these solutions boundaries parallel to x-axis are considered as insulated boundaries which reduces the problem to a one dimensional one. Error propagation obtained for problem one is shown in Figure (2) comparatively. In this comparison SFD implies standard finite difference discretization with dt = 0.01, SASR implies semianalytic SR-BEM solution with zero and one internal points and SADR implies semianalytic DR-BEM solution with zero internal points. As can be seen from these results SASR and SADR results are more accurate than SFD solution. When compared with SASR solution, SADR solution is more accurate at least at initial times. This is mainly due to domain integration which needs to be completed in SASR, whereas SADR is a domain integration free process. Error propagation in problem two is shown in Figure (3) with similar trends. In both of these applications the error is defined in terms of the Euclidian norm definition where numerical results are compared with analytical results.

CONCLUSION

A new numerical time integration process is described for the solution of time dependent problems which combines the advantages of boundary element procedures with analytic time integration processes. Results obtained are encouraging. Numerical results obtained for the two test problems and the comparison of these results with other numerical procedures demonstrate the accuracy and efficiency of the proposed technique.

REFERENCES

1. Aral, M. M. and Tang, Y., A Boundary Only Procedure for Time-Dependent Diffusion Equation, Applied Math. Modelling, Vol. 12, pp. 610-618, 1988.
2. Aral, M. M. and Tang, Y., A New Boundary Element Formulation for Time Dependent Confined and Unconfined Aquifer Problems, Water Resources Research, Vol. 24, No. 6, pp. 831-842, 1988.
3. Aral, M. M. and Tang, Y., A Boundary Only Procedure for Transient Transport Problems with or without First-Order Chemical Reaction, Applied Math. Modelling, Vol. 13, 1989.
4. Chang, Y. P., Kang, C. S., and Chen, D. J., The Use of a Fundamental Greens's Functions for the Solution of Problems of Heat Conduction in Anisotropic Media, Int. J. Heat and Mass Transfer, Vol. 16, pp. 1905-1918, 1973.
5. Curran, D. A. S., Cross, M., and Lewis, B. A., Solution of Parabolic Partial Differential Equations by the Boundary Element Method Using Discretization in Time, Applied Math. Modelling, Vol. 4, pp. 398-406, 1980.
6. Rizzo, F. J., and Shippy, D. J., A Method of Solution for Certain Problems of Transient Heat Conduction, AIAA, Vol. 8, No. 11, pp. 2004-2009, 1970.
7. Umari, A. M. and Gorelick, S. M., The Problem of Complex Eigensystems in the Semianalytic Solution for Advancement of Time in Solute Transport Simulations: A New Method Using Real Arithmetic, Water Resources Research, Vol. 22, No. 7, pp. 1149-1154, 1986.
8. Wrobel, L. C., Brebbia, C. A., and Nardini, D., The Dual Reciprocity Boundary Element Formulation for Transient Heat Conduction, Proc. VI Int. Conf. on Finite Elements in Water Resources, Lisbon, Portugal, pp. 801-812, 1986.

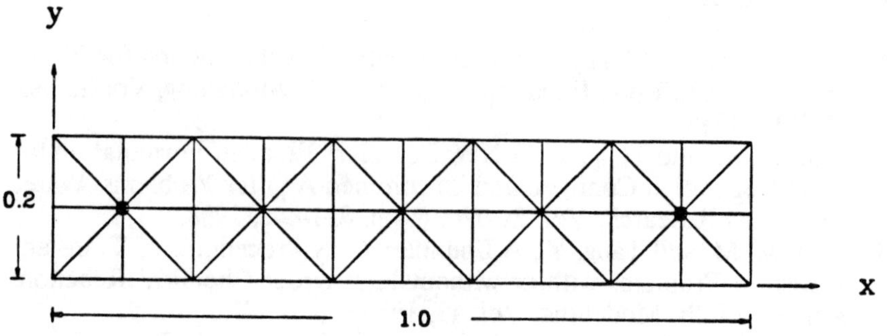

Figure 1. Idealization of the Solution Domain

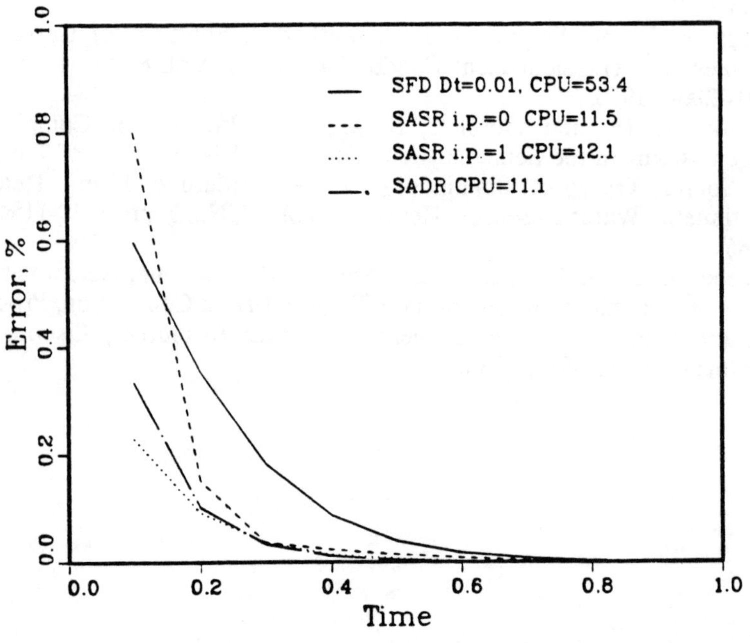

Figure 2. Error Propagation in Problem One

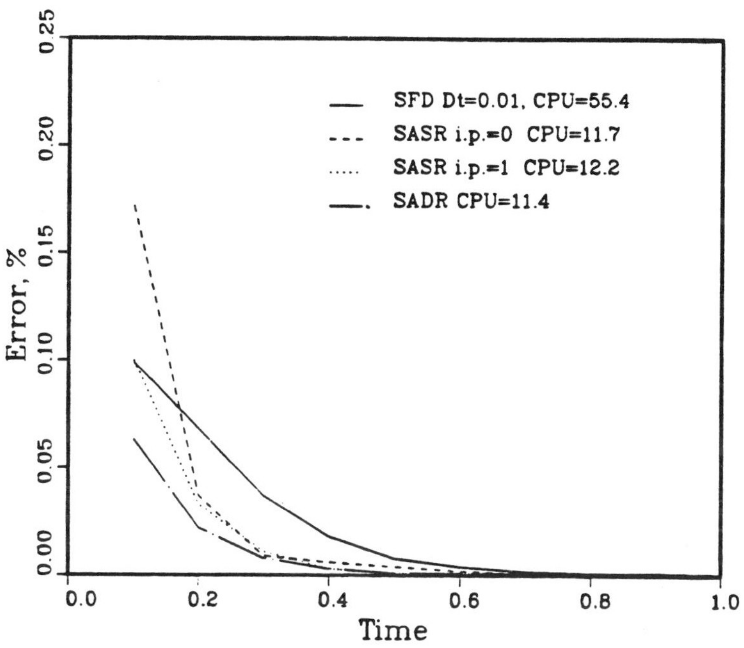

Figure 3. Error Propagation in Problem Two

5000 # 86732 Engineering "B"